The Handbook
of Environmental Chemistry

Volume 5   Water Pollution
Part C

O. Hutzinger
Editor-in-Chief

Springer-Verlag Berlin Heidelberg GmbH

Quality and Treatment of Drinking Water II

Volume Editor: J. Hrubec

With contributions by
R. A. Baumann, I. R. Falconer, J. K. Fawell,
J. Hoigné, H. Horth, J. Hrubec,
F. X. R. van Leeuwen, C. Rav-Acha,
P. van Zoonen

 Springer

Environmental chemistry is a rather young and interdisciplinary field of science. Its aim is a complete description of the environment and of transformations occuring on a local or global scale. Environmental chemistry also gives an account of the impact of man's activities on the natural environment by describing observed changes.

"The Handbook of Environmental Chemistry" provides the compilation of today's knowledge. Contributions are written by leading experts with practical experience in their fields. The Handbook will grow with the increase in our scientific understandig and should provide a valuable source not only for scientists, but also for environmental managers and decision makers.

ISSN 1433-6863
ISBN 978-3-662-14774-0

Library of Congress Cataloging-in-Publication Data
The Natural environment and the biogeochemical cycles /
with contributions by P. Craig ... [et al.].
v. <A-F > : ill. ; 25 cm. -- (The Handbook of environmental chemistry :
v. 1) Includes bibliographical refereces and indexes.
ISBN 978-3-662-14774-0   ISBN 978-3-540-68089-5 (eBook)
DOI 10.1007/978-3-540-68089-5
1. Biogeochemical cycles. 2. Environmental chemistry.
I. Craig. P. J., 1944- . II. Series.
QD31. H335 vol. 1   [QH344]   628.5 s

© Springer-Verlag Berlin Heidelberg 1998
Originally published by Springer-Verlag Berlin Heidelberg New York in 1998
Softcover reprint of the hardcover 1st edition 1998

The use of general descriptive names, registered names, trademarks, etc. in this publication does not imply, even in the absence of a specific statement, that such names are exempt from the relevant protective laws and regulations and therefore free for general use.

Product liability: The publisher cannot guarantee the accuracy of any information about dosage and application contained in this book. In every individual case the user must check such information by consulting the relevant literature.

The instructions given for the practical carrying-out of HPLC steps and preparatory investigations do not absolve the reader from being responsible for safety precautions. Liability is not accepted by the author.

Production Editor: ProduServ GmbH Verlagsservice, Berlin
Cover design: E. Kirchner, Springer-Verlag
Typesetting: Fotosatz-Service Köhler OHG, Würzburg
SPIN: 10554920   52/3020 - 5 4 3 2 1 0 - Printed on acid-free paper

Advisory Board

Editor-in-Chief

Prof. Dr. Otto Hutzinger

Universität Bayreuth
Lehrstuhl für Ökologische Chemie
und Geochemie
Postfach 10 12 51
D-95440 Bayreuth, Germany
E-mail: otto.hutzinger@uni-bayreuth.de

Volume Editor

Dr. Jiři Hrubec

Laboratory for Water and
Drinking Water Research
National Institute of Public Health
and Environmental Protection (rivm)
Antonie van Leeuwenhoeklaan 9
NL-3720 BA Bilthoven, The Netherlands
E-mail: Jiri.Hrubec@rivm.nl

Preface

Environmental Chemistry is a relatively young science. Interest in this subject, however, is growing very rapidly and, although no agreement has been reached as yet about the exact content and limits of this interdisciplinary discipline, there appears to be increasing interest in seeing environmental topics which are based on chemistry embodied in this subject. One of the first objectives of Environmental Chemistry must be the study of the environment and of natural chemical processes which occur in the environment. A major purpose of this series on Environmental Chemistry, therefore, is to present a reasonably uniform view of various aspects of the chemistry of the environment and chemical reactions occurring in the environment.

The industrial activities of man have given a new dimension to Environmental Chemistry. We have now synthesized and described over five million chemical compounds and chemical industry produces about hundred and fifty million tons of synthetic chemicals annually. We ship billions of tons of oil per year and through mining operations and other geophysical modifications, large quantities of inorganic and organic materials are released from their natural deposits. Cities and metropolitan areas of up to 15 million inhabitants produce large quantities of waste in relatively small and confined areas. Much of the chemical products and waste products of modern society are released into the environment either during production, storage, transport, use or ultimate disposal. These released materials participate in natural cycles and reactions and frequently lead to interference and disturbance of natural systems.

Environmental Chemistry is concerned with reactions in the environment. It is about distribution and equilibria between environmental compartments. It is about reactions, pathways, thermodynamics and kinetics. An important purpose of this Handbook, is to aid understanding of the basic distribution and chemical reaction processes which occur in the environment.

Laws regulating toxic substances in various countries are designed to assess and control risk of chemicals to man and his environment. Science can contribute in two areas to this assessment; firstly in the area of toxicology and secondly in the area of chemical exposure. The available concentration ("environmental exposure concentration") depends on the fate of chemical compounds in the environment and thus their distribution and reaction behaviour in the environment. One very important contribution of Environmental Chemistry to the above mentioned toxic substances laws is to develop laboratory test methods, or mathematical correlations and models that predict the environmental fate of new chemical compounds. The third purpose of this Handbook is to help in the basic understanding and development of such test methods and models.

The last explicit purpose of the Handbook is to present, in concise form, the most important properties relating to environmental chemistry and hazard assessment for the most important series of chemical compounds.

At the moment three volumes of the Handbook are planned. Volume 1 deals with the natural environment and the biogeochemical cycles therein, including some background information such as energetics and ecology. Volume 2 is concerned with reactions and processes in the environment and deals with physical factors such as transport and adsorption, and chemical, photochemical and biochemical reactions in the environment, as well as some aspects of pharmacokinetics and metabolism within organisms. Volume 3 deals with anthropogenic compounds, their chemical backgrounds, production methods and information about their use, their environmental behaviour, analytical methodology and some important aspects of their toxic effects. The material for volume 1, 2 and 3 was each more than could easily be fitted into a single volume, and for this reason, as well as for the purpose of rapid publication of available manuscripts, all three volumes were divided in the parts A and B. Part A of all three volumes is now being published and the second part of each of these volumes should appear about six months thereafter. Publisher and editor hope to keep materials of the volumes one to three up to date and to extend coverage in the subject areas by publishing further parts in the future. Plans also exist for volumes dealing with different subject matter such as analysis, chemical technology and toxicology, and readers are encouraged to offer suggestions and advice as to future editions of "The Handbook of Environmental Chemistry".

Most chapters in the Handbook are written to a fairly advanced level and should be of interest to the graduate student and practising scientist. I also hope that the subject matter treated will be of interest to people outside chemistry and to scientists in industry as well as government and regulatory bodies. It would be very satisfying for me to see the books used as a basis for developing graduate courses in Environmental Chemistry.

Due to the breadth of the subject matter, it was not easy to edit this Handbook. Specialists had to be found in quite different areas of science who were willing to contribute a chapter within the prescribed schedule. It is with great satisfaction that I thank all 52 authors from 8 countries for their understanding and for devoting their time to this effort. Special thanks are due to Dr. F. Boschke of Springer for his advice and discussions throughout all stages of preparation of the Handbook. Mrs. A. Heinrich of Springer has significantly contributed to the technical development of the book through her conscientious and efficient work. Finally I like to thank my family, students and colleagues for being so patient with me during several critical phases of preparation for the Handbook, and to some colleagues and the secretaries for technical help.

I consider it a privilege to see my chosen subject grow. My interest in Environmental Chemistry dates back to my early college days in Vienna. I received significant impulses during my postdoctoral period at the University of California and my interest slowly developed during my time with the National Research

Council of Canada, before I could devote my full time of Environmental Chemistry, here in Amsterdam. I hope this Handbook may help deepen the interest of other scientists in this subject.

Amsterdam, May 1980   *O. Hutzinger*

Seventeen years have now passed since the appearance of the first volumes of the Handbook. Although the basic concept has remained the same some changes and adjustments were necessary.

Some years ago publishers and editor agreed to expand the Handbook by two new open-ended volume series: Air Pollution and Water Pollution. These broad topics could not be fitted easily into the headings of the first three volumes. All five volumes series are integrated through the choice of topics and by a system of cross referencing.

The outline of the Handbook is thus as follows:

1. The Natural Environment and the Biochemical Cycles,
2. Reactions and Processes,
3. Anthropogenic Compounds,
4. Air Pollution,
5. Water Pollution.

Rapid developments in Environmental Chemistry and the increasing breadth of the subject matter covered made it necessary to establish volume-editors. Each subject is not supervised by specialists in their respective fields.

A recent development is the 'Super Index', a subject index covering chapters of all published volumes, which will soon be available via the Springer Home-page http://www.springer.de or http://www.springer-ny.com or http://Link. springer.de.

With books in press and in preparation we have now published well over 30 volumes. Authors, volume-editors and editor-in-chief are rewarded by the broad acceptance of the 'Handbook' in the scientific community.

May 1997   *Otto Hutzinger*

Contents

Introduction

In the last decades contamination of drinking water and growing public concern about the health risks of contaminants have received much publicity and initialized many research efforts, as well as political and legal activities.

The majority of the recent problems related to drinking water contamination, associated with pollution of surface and ground water resources and with the formation of reaction by-products resulting from the use of disinfectants and oxidants in drinking water treatment, is closely connected with the rapid advances in analytical techniques. The modern analytical methods have resulted in the identification of a huge number of chemical compounds and microbial pollutants since the early seventies. Continuing discoveries of new drinking water pollutants and related health hazards have had a shocking effect on the public. For the professional community they have created a multitude of unknown factors and uncertainties concerning toxicological, technological and regulatory aspects.

One of the major issues related to drinking water contamination is the assessment of the health hazards and associated risk comparisons, priority settings and risk management. The health hazard assessesment plays an important role in the evaluation of the overall relevance of the problem and is one of the principal factors in the formulation of research needs. A specific feature of health hazards related to drinking water contamination constitutes a dilemma of "competing risk", leading to reduction of a "target risk" and simultaneously creating other kinds of risks. A well known example is the use of chemical disinfectants for elimination of microbial risk, resulting in an increase of health risks from the formation of reaction by-products and vice versa. As a result, reduction of risk from formation of by-products by restrictive measures in the application of chemical can result in an increase of microbial risk.

Health risk assessment has a decisive influence on the setting of national and international quality standards and directives. Due to the current limited state of scientific knowledge and the complexity of political and social reality the quality standards have only a temporary character and therefore constitute an unstable, but nevertheless the only available rational basis for the formulation of technological and technical goals and objectives.

As far as treatment of drinking water is concerned, since 1974, when the formation of trihalomethanes by chlorination was discovered, chlorination by-products are the major research topic. A large number of studies on identification of the reaction by-products of chlorination and on their toxicological effects has provided convincing reasons for avoiding the use of chlorine in drinking water treatment and for the use of alternative disinfection methods.

However, much less information is available on the consequences of the application of alternatives for chlorine, such as ozone and chlorine dioxide. Still, insufficient evidence exists that the reaction by-products of alternative disinfectants and oxidants are less hazardous than those of chlorine. An important warning, which can be learned from the research on alternative disinfectants for drinking water treatment, is the fact that the application of any "transformation" process in drinking water treatment introduces a high risk of formation of by-products, which are currently largely unidentifiable and have unknown health effects. Clearly preference should be given to "real removal" processes, such as aeration, adsorption on activated carbon and membrane separation. Considerable progress has been made recently in the understanding and in the practical application of these processes.

The most serious threat for drinking water quality is posed by the pollution of drinking water resources. As far as surface water is concerned it is caused by anthropogenic compounds, by pathogenic microorganisms and by pollutants related to eutrophication, such as odor and taste compounds and algae toxins.

Ground water, traditionally considered as the safest drinking water source, has been threatened more and more by waste dumping, by nitrate and pesticides, resulting from agricultural activities and from air pollution.

Finally, still more attention is being given to the quality deterioration of drinking water during transportation and storage as a result of material corrosion and biological activity promoted by the presence of biodegradable compounds.

The handbook does not attempt to be an exhaustive review of such a vast and complex subject as drinking water quality, but it is meant to give an overview of the developments in key areas related to chemical contamination, with special attention to organic micropollutants.

The two volumes of the handbook dealing with the subject of drinking water are organized as follows.

The published first volume (Volume 5, Part B) has principally addressed:

- the latest developments in quality regulation;
- the role of biological processes in degradation of organic micropollutants and in control of biological instability of drinking water;
- significance of biological stability of drinking water;
- control of organic micropollutants by adsorption on activated carbon;
- origin and removal of odors and tastes.

This second volume focusses on the following topics:

- toxicity tests for assessing drinking water quality;
- toxicological approaches for developing drinking water standards;
- analysis of organic micropollutants;
- algal toxins and human health;
- quality changes due to application of ozone and chlorine dioxide.

From the important drinking water quality issues, the volumes do not address microbiological pollution because of the scope of the Handbook. From the chemical issues, the principal topic of the reaction by-products of chlorination is not addressed, mainly because it is covered in great detail in a number of other publications. One of the basic aspects of the chlorination problem – health risks of chlorinated drinking water – has already been reviewed in the Handbook elsewhere (see Craun GF, vol 5, part A, p 1).

Jíři Hrubec

Toxicological Approaches for Developing Drinking Water Standards

F. X. Rolaf van Leeuwen

WHO European Centre for Environment and Health, Bilthoven Division, P.O. BOX 10, 3730-AA De Bilt, The Netherlands
E-mail: eceh@who.nl

Drinking-water guidelines and standards are an important tool to protect public health and provide regulator agencies and drinking-water producers with guidance on the production of adequate and safe drinking water. The potential hazard of chemicals polluting drinking water or resulting from the treatment of water are widely understood but often overestimated in relation to microbiological hazards. This chapter will provide some insight into the toxicological evaluation which forms the basis for the derivation of guidelines for drinking-water quality as set by the World Health Organization and will touch upon some differences between health-based guidelines and national standards. Special attention will be given to the toxicological impact of chemicals present in drinking water such as inorganic pollutants, pesticides and disinfectants and disinfectant by-products.

Keywords: chemical pollutants, drinking water quality, guidelines, health risk evaluation.

Contents

List of Abbreviations

BW   body weight
C   concentration

The Handbook of Environmental Chemistry Vol. 5 Part C
Quality and Treatment of Drinking Water II (ed. by J. Hrubec)
© Springer-Verlag Berlin Heidelberg 1998

2,4 D   2,4 dichlorophenoxyacetic acid
2,4 DB   4-(2,4 dichlorophenoxy)butyric acid
DBP   disinfection by-product
DNA   deoxyribonucleic acid
EU   European Union
FAO   Food and Agricultural Organization
GV   guideline value
LOAEL   lowest observed adverse effect level
NOAEL   no observed adverse effect level
OECD   Organization for Economic Co-operation and Development
ppb   parts per billion
TDI   tolerable daily intake
TWI   tolerable weekly intake
USEPA   United States Environmental Protection Agency
WHO   World Health Organization

1
Introduction

Drinking water is a basic requirement for sustaining life. Since the whole
human population needs drinking water, the provision of a safe water supply,
therefore, is a high priority issue for safeguarding the health and well being of
humans. The production of adequate and safe drinking water is the most
important factor contributing to a decrease in mortality and morbidity in
developing countries. As reported by the World Health Organization [1], nearly
half of the population in these countries suffers from health problems associated with lack of drinking water or the presence of microbiologically contaminated water. Traditionally microbiological contamination and quality has
been the main concern. Although this concern has not been reduced in recent
years the attention of the general public and health officials on the importance
of chemical quality has increased with the increase of our knowledge on the
hazards of exposure to chemical substances. It should be understood, however,
that the public health risk from microbiological contamination differs from that
of chemical pollutants. The risk of microbiological contamination usually is
acute and wide spread, whereas health risk associated with chemical contamination usually concerns effects following low level exposure for a prolonged
period.

One of the key issues in drinking water production is how to rid drinking
water of potentially dangerous microorganisms and chemicals without introducing new hazards that might pose new and different threats to human
health. This chapter will concentrate on chemical risk assessment and on
the derivation of guidelines for drinking water quality for chemical compounds to safeguard public health from a chemical/toxicological point of view.

The potential for exposure of the human population to chemicals through
drinking water is enormous and we must assume that there is a potential for all
members of the population, including potentially high risk groups such as

young children and people with poor health, to use drinking water that might contain possible hazardous contaminants. Therefore, strict quality requirements should be set to protect public health.

There are many sources of contamination of drinking water. Broadly they can be divided into two categories: contaminants originating from ground and surface water, the sources for drinking water production, and contaminants used or formed during the treatment and distribution of drinking water. Surface and ground water are vulnerable to spills and accidental contamination from industrial and agricultural chemicals but unfortunately surface water is also a convenient disposal route for many wastes. Contaminants in ground and surface water will range from natural substances leached from soil, runoff from agricultural activities, controlled discharge from sewage treatment works and industrial plants, and uncontrolled discharges or leakages from landfill sites and from chemical accidents or disasters.

Naturally occurring contaminants are predominantly formed by inorganic compounds such as arsenic and manganese which are derived from natural mineral formations. Organic compounds, pesticides, disinfectants and disinfectant by-products are introduced by human activity.

So it can be said that drinking water quality does not only depend on the quality of the water resources but also on the type and quality of the processes used by the drinking water industry for the treatment and distribution of water.

To assure consumers that drinking water is safe and can be consumed without any risk, guidelines or standards have to be set, giving maximum allowable concentrations for compounds in drinking water below which no significant health risk is encountered. This chapter will focus primarily on the toxicological procedures used by the World Health Organization to derive guideline values for chemical compounds in drinking water, and will touch upon some critical differences in the nature of guidelines and national standards.

Table 1. Selected chemical substances for the first addendum to the WHO-GDWQ

Inorganics	Organics	Pesticides	Disinfectants
aluminium	cyanotoxins	bentazone	chloroform
boron	EDTA	carbofuran	dichloroisocyanurate
copper	PAHs	cyanizine	
nickel	trichloroethylene	2,4-D	
NO_2/NO_3		1,2 dichloropropane	
uranium		diquat	
		ethylene dibromide	
		glyphosate	
		pentachlorophenol	
		terbuthylazine	

2
Health Related Guidelines or Standards

In establishing Guidelines for drinking-water to protect public health, the World Health Organization [2] applied the following definition of a guideline for drinking water quality:

"A guideline value represents the concentration of a constituent that does not result in any significant risk to health of the consumer over a lifetime of consumption."

Although this guideline value provides a maximum level of a contaminant that may not cause any public health concern even following lifetime exposure, it does not implicate a "green light" for pollution of drinking water to the recommended guideline level. It must be recognized that because water is essential to sustain life a continuous effort should be made to maintain water quality at the highest possible level.

WHO drinking-water guidelines (WHO-GDWQ) are health based guidelines derived on the basis of internationally agreed procedures for risk assessment. They are not standards per se, and therefore should not be considered as mandatory limits, but are intended to be used by national authorities as a basis for national standards on water quality. In setting national, legally binding standards, several important issues must be considered such as technical feasibility, monitoring facilities, geographical and environmental situation, and socio-economic and cultural conditions. The setting of standards might be influenced by national priorities and economic considerations and thus the conclusion on whether the health benefit of setting a specific standard justifies the costs involved is a matter for each individual country. In this way standards can be tailored to the local situation and can be implemented and controlled in the most effective manner. For detailed information on the practical consequences of implementation of standards and the required control policies the reader is referred to the chapter by Van Dijk-Looijaard [3] in this volume. Although the varying geographical situations and national needs might thus lead to different standards for drinking water quality in different countries, it must never be allowed that these considerations could endanger public health.

The first edition of the WHO Guidelines for drinking-water quality [4] was published in 1984 and the second edition [2] in 1993. These Guidelines were widely accepted and recognized by numerous Member States as being of high value. Due to the emergence of new scientific data and changes in priorities it was agreed that there would be a continuing process of updating of the Guidelines and that addenda to the second edition would be published containing information on the health risk of selected chemical (and microbiological) contaminants. At present, this revision of WHO Guidelines for drinking-water quality is well underway, and publication of the first addendum including the compounds indicated in Table 1, is expected by the end of 1997 [5].

3
Toxicological Basis for Evaluation

Toxicology is the study of the effects of chemical compounds on living organisms. In toxicology, three different fields of interest can be distinguished: descriptive toxicology, mechanistic toxicology and regulatory toxicology.

Descriptive toxicology characterizes the adverse effects of chemicals particularly obtained in studies with experimental animals, or in clinical or epidemiological studies. It is the aim of these studies to identify the potential hazard for human health associated with exposure to chemicals.

Mechanistic toxicology tries to unravel the way in which chemicals exert their adverse effects in living organisms. For this goal studies with laboratory animals as well as various in vitro models are in use. Based on the information obtained in these studies also structure-activity relationships for toxic compounds can be established.

Regulatory toxicology aims at the assessment of the risk for human health posed by chemicals and their breakdown products, based on the evaluation of all available information, leading to an expert judgment whether the chemical compounds can be used safely and thus will not pose any unacceptable health risk to the human population. This process of preventive risk assessment often culminates in the derivation of safe exposure levels for chemical compounds for the general population. In the field of drinking water quality these levels are reflected by drinking water guidelines or standards.

Regulatory toxicology is based on two main principles in hazard assessment: knowledge of the fate of the compound (pharmacokinetic studies) and knowledge on the specific effects of the compound (toxicity studies).

Kinetic experiments give insight into the adsorption of a compound into an organism, the distribution throughout the body, the metabolic processes undergone by the compound and the possible formation of more harmful or harmless breakdown products (metabolites), and the rate of excretion of the compound or its metabolites from the body via urine or faeces. In general the manifestation of toxic effects does not depend directly on the administered (external) dose but rather on the concentration of the compounds or its metabolites in the organs where the effects occur (target organs).

For the evaluation of the toxic effects of chemicals regulatory toxicologists often have to rely on animal studies only. In these cases it is of utmost importance that these animal studies fulfil some basic requirements which facilitate the comparison of animal data with the human situation. The first requirement is that the effects observed in laboratory animals are applicable to humans. Or, in other words, that the selected animal species is representative for humans. Secondly the animal studies must fulfil some minimum requirements with respect to the study protocol (e.g. type of test species, number of animals used, duration of the study, parameters examined). Nowadays most of the studies provided by industrial laboratories for evaluation are in compliance with OECD protocols laying down the minimum requirements for several types of toxicity studies [6]. Finally the route of administration in the animal studies should preferentially be identical to the route of administration which is most relevant for the human risk assessment. In the case of drinking water regulation, this means that the oral route is preferred.

The spectrum of adverse effects produced by chemical compounds is huge and ranges from relatively harmless to life threatening. The seriousness of the effects is determined by a number of features: the onset of the effects after exposure (acute or chronic), nature of the effects (functional or structural), whether the effects are permanent or transient (irreversible or reversible). For the evaluation of the toxic effects of chemicals in drinking water the main emphasis is placed on studies with chronic exposure (life span) or at least subchronic (about 1/10 life span) because concentrations of contaminants are relatively low and consumption of drinking water takes place over our entire lifetimes. These studies aim at the characterization of biochemical, functional and structural changes in a large number of target tissues, such as liver, kidneys, lung, brain, reproductive organs, bone marrow, and blood. In addition to these parameters information on potential genetic defects is of crucial qualitative importance for the hazard assessment. This information is collected from mutagenicity and carcinogenicity studies.

Mutagenicity is the ability of a compound to irreversibly affect DNA so that the changes are retained after division of cells (somatic cells) or will be passed to subsequent generations (germinal cells). In general these events can occur through gene mutation or chromosomal aberrations. Chemicals exerting this type of activity are referred to as genotoxic, and it is assumed that these chemicals may cause cancer by an initial mutagenic event.

Cancer is defined as unrestrained cell proliferation. Groups of rapidly dividing cells that lack the ability to perform the same function as unaffected neighboring cells are called tumors or neoplasms. Genotoxic carcinogens are thought to cause cancer by mutational activation of naturally existing pro-oncogenes, coding for proteins that increase the response of cells to endogenous molecules stimulating cell growth or controlling cell differentation.

For genotoxic carcinogens it is assumed that there is a theoretical probability that a molecule of these substances is capable of interacting with DNA and so producing a genetically transformed cell that subsequently proliferates to form a tumor. This theory is referred to as "the one hit theory", and implies that no threshold of exposure exists below which such a carcinogenic effect may not exist.

A second group of carcinogens called tumor promoters or non-genotoxic carcinogens produce tumors through a mechanism that does not involve DNA damage and are thought to be incapable of producing cancer by themselves but rather enhance the formation of tumors following an initial mutational event. For compounds causing cancer through such an indirect mechanism, it is assumed that, in contrast to genotoxic carcinogens, a threshold exist below which the probability of eliciting a carcinogenic response in an individual exposed to such compounds is zero, or at least negligible. A phenomenon that is also associated with the toxic responses of non-carcinogenic toxic compounds.

This distinction in compounds exerting their toxic effect either through a threshold or a non-threshold mechanism is crucial in the health risk evaluation of chemicals and thus also in the health risk evaluation of drinking water contaminants because it leads to two distinct approaches in setting safe or acceptable risk levels for exposure to chemicals: the Tolerable Daily Intake (TDI) approach or a mathematical low-dose risk extrapolation resulting in a unit risk. The impact of this important consideration will be illustrated in the next section.

4
Derivation of Guidelines

For the health risk assessment of chemicals, generally two sources of information are available: toxicity studies using laboratory animals and studies in human populations. Both sources have their own limitations. In animal studies, a relatively small number of animals is usually used and the doses administered, therefore, are relatively high in order to exert an observable toxic response. The results of these studies need to be extrapolated to the human situation with, in general, lower levels of exposure. When several studies in various animal species are available one should select the best conducted studies with the most comparable species and the most relevant effects for the human risk assessment.

Studies in humans are mostly based on data collected from occupational settings often with simultaneous exposure to several compounds. Therefore these studies often lack reliable quantitative exposure information. For clinical case control studies, this is not the case but their number is very limited and, in general, the number of volunteers involved in these studies is relatively small. In order to derive guideline values that protect human health, expert judgment is exercised to select the most appropriate information from the range of studies that is available.

4.1
Tolerable Daily Intake (TDI)

The basis for the guideline derivation for compounds exerting their effects through a threshold mechanism is the Tolerable Daily Intake (TDI). The TDI is an estimate of the amount of a chemical present in food or drinking water that can be ingested daily over a lifetime, without appreciable health risk [7]. This TDI is expressed in mg or μg of the compound per kg body weight. Ideally the TDI is derived from a so called No Observed Adverse Effect Level (NOAEL) established in toxicity studies. This is the highest dose in toxicity studies at which no detectable adverse changes in organ function or morphology, growth, developmental parameters, or life span of the test species are detected [8]. In some instances, however, such a level without effect is not available and the TDI has to be based on the lowest dose level at which such effects have been observed (LOAEL). Preferentially the NOAEL or the LOAEL is based on long-term toxicity studies. This NOAEL or LOAEL is then divided by an appropriate uncertainty factor to correct for extrapolation from animal to human data, and for (in)adequacy of the database. In the derivation of guideline values for drinking water quality published by the WHO [2] the following uncertainty factors were applied by the WHO:

- extrapolation from animals to humans 1 – 10
- individual variation 1 – 10
- (in)adequacy of the database 1 – 10
- nature and severity of effect 1 – 10

The database might be inadequate when a LOAEL has to be used for derivation of the TDI instead of a NOAEL, or when studies have to be considered with shorter duration (subchronic) than actually desirable (chronic). When the toxicological endpoint considered comprises effects directly related with a carcinogenic potential or in the case of structural malformations of feuses in reproductive toxicity tests, an additional uncertainty factor might be warranted.

In general the default value for these uncertainty factors is 10, however, lower figures might be used based on the information obtained from the available dataset. The WHO [2] concluded that the total uncertainty factor should not exceed 10 000. If the expert judgment leads to a higher uncertainty factor, it was decided that in such cases, the TDI is so imprecise that it lacks any physiological significance. It should be noted that the selection of uncertainty factors is a very important step in the derivation of the TDI (and thus of the guideline value) as they can make a considerable difference in the values set and it was therefore expressed by the WHO that the selection of uncertainty factors should be clarified where possible [2].

For chemicals for which it is assumed that a threshold exists for the occurrence of adverse toxic effects, a TDI is established as follows:

$$TDI = \frac{NOAEL \text{ or } LOAEL}{uncertainty \text{ factor}}$$

In setting guidelines or standards for drinking water quality, it has to be realized that exposure of the human population to chemical compounds occurs not through drinking water alone. In many cases, intake through drinking water is lower in comparison to intake via food or exposure via air. Therefore, the WHO [2] allocated only a fraction of the TDI to drinking water to compensate for exposure through other routes and the ensure that the total intake of the human population from all sources does not exceed the TDI. When exposure data from different sources were available based on mean levels in food, air and drinking water, these data were used in the derivation of the guideline value. For some pesticides for instance with high potential exposure through food, an allocation of 1% was applied. For compounds where exposure only results from their use in drinking water treatment or distribution (disinfectants and disinfectant by-products) sometimes high allocation percentages (e. g. 80%) were applied. In cases were appropriate exposure data were lacking a default allocation of 10% was used.

The guideline value (GL) is derived from the TDI as follows:

$$GL = \frac{TDI \times BW \times P}{C}$$

Where BW = body weight (60 kg for adults, 10 kg for children, 5 kg for infants), P = fraction of the TDI allocated to drinking water, and C = daily water consumption (2 L for adults, 1 L for children, 0.75 L for infants).

It should be emphasized here that the allocation factors used by the WHO in the derivation of the guidelines for drinking-water quality are based on overall exposure scenarios and that local situations might differ considerably. This implies that the guidelines might not be necessarily applicable to all areas or regions, and that authorities should be encouraged to derive guidelines or standards by using allocation figures that are tailored to the local circumstances [2].

4.2
Mathematical Low-Dose Risk Extrapolation

For compounds considered to be genotoxic carcinogens, it is assumed that no threshold exists for the carcinogenic potential, and that there is a probability of harm at any level. Therefore the application of the TDI-approach is considered inappropriate. For the derivation of a guideline, mathematical low-dose risk extrapolation is applied. This extrapolation is usually based on long-term animal studies, although in some cases information on the carcinogenicity of the compound(s) in humans is available, mostly from occupational exposure. If the risk extrapolation has to be carried out on animal data the process is actually based on two assumption: 1) humans are as sensitive as the most sensitive test species, and 2) the mechanism by which the chemical causes cancer is assumed to be consistent with the model used for human risk extrapolation.

There are several mathematical models such as probit, logit, Weibul, one hit, and multistage models. There are many assumptions built into these models and at best they could be considered as simplified probablistic representations of complex biological processes that err on the side of caution. They make use of relatively high dose levels to compensate for the relatively small number of test animals, and they do not properly take into account important biological processes such as DNA repair, immunological protection mechanisms and pharmacokinetics. It is well known that, although these models give similar risk estimates at high dose levels they give highly diverging risk estimates at lower level. An example of this is shown in Fig. 1 illustrating the results presented by Cotruvo [9] of the risk extrapolation of trichloroethylene. It can be seen that the risk estimates obtained with the various models differ by several orders of magnitude.

The WHO [2] has chosen the linearized multistage model also favored by the USEPA as the model to derive guidelines for carcinogenic drinking water pollutants, realizing that this model is rather conservative and therefore will usually overestimate the risk. In addition, the risk is presented as the upper boundary 95% confidence limit, which means that there is only a 5% chance that the calculated risk will be exceeded at the dose given, while the lower 95% confidence range usually encompass a zero risk.

Although it was recognized by the WHO that differences in metabolic rates between test animals and humans are better correlated with the ratio of body

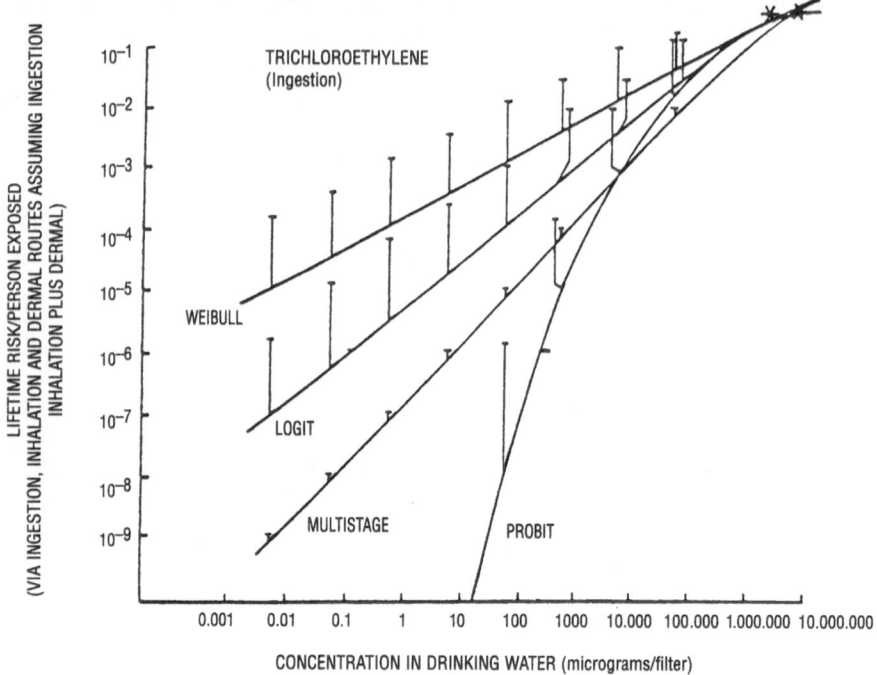

Fig. 1. Sample risk extrapolations for trichloroethylene using four different models [9]

surface area than with body weight, it was decided, in line with the conclusion of Crump et al. [10] that all measures of dose except dose rate per unit body weight tend to overestimate the risk for humans. Therefore, the guidelines for carcinogenic compounds are expressed on a body weight basis. For compounds considered to be genotoxic carcinogens, guideline values are presented as the concentration in drinking water associated with an excess lifetime cancer risk of one case in a population of 100 000.

5
Inorganic Pollutants

Inorganic compounds form the largest proportion of chemicals in drinking water. They are usually present as a consequence of their natural occurrence and can be considered as an integral part of drinking water sources. In general, the concentration of inorganic compounds in raw water is much higher than that of organic pollutants, although the concentrations can differ considerably due to the differences between groundwater and surface water.

The health risk assessment essential for the derivation of drinking-water guidelines for inorganic compounds is hampered by the limited database on the toxicity of these compounds. In particular information of health effects following oral administration is scarce. In addition, several of the inorganic contaminants have beneficial as well as adverse effects. A number of these elements such as copper, iron, manganese, molybdenum, selenium, and zinc are essential elements in human nutrition. However, because it is assumed that the "essential" intake of these elements is primarily covered by food, no attempt have been made to set minimum desirable levels of these elements in drinking water.

Some inorganic pollutants, however, are primarily the result of human activity. Well-known examples are lead, from lead piping and plumbing, and nitrate, from intensive agricultural activity.

5.1
Lead

The lead concentration in drinking water depends on the occurrence of lead piping and plumbing and the hardness, temperature and pH of the water. Lead is a cumulative toxicant that can cause neurological and behavioral effects, particularly in young children. The WHO/FAO Joint Expert Committee on Food Additives [11] has established a tolerable weekly intake (TWI) of 25 µg/kg body weight for infants and children on the basis that there should be no accumulation of the body burden of lead. This figure has been derived from metabolic studies showing that blood lead levels in infants were not increased at an intake of 3–4 µg/kg body weight per day. In deriving a guideline value for lead in drinking water, the WHO recognized that infants are the most sensitive members of the population and that setting a guideline value to protect them would also be protective for other age groups [2]. By allocating 50 % of this TWI to drinking water for a 5 kg (bottle fed) infant consuming 0.75 l water per day, a guideline value of 0.01 mg/l was established. In many countries, programmes have been set up to remove plumbing and service connections containing lead in order to reduce the lead content of drinking water, particularly in inner cities. However, it should be recognized that this is a very costly and time-consuming operation.

5.2
Nitrate

Another contaminant that attracts public attention is nitrate. The naturally occurring levels in ground- and surface water are usually several milligrams per liter, however, as a consequence of the use of fertilizers and the spread of manure over agricultural land, concentrations can increase to several hundreds of milligrams per liter. Due to the high nitrate levels in the upper layers of agricultural land it is expected that nitrate levels in groundwater will continue to increase in the near future.

The main public health concern over nitrate relates to adverse effects in bottle-fed infants. It is well known that nitrate is reduced in the body to nitrite and causes oxidation of haemoglobin to methaemoglobin and thus reduces the oxygen binding capacity of this protein. This might lead to the so called "blue-baby syndrome" in cases were the methaemoglobin concentration reaches 10 % or more, compared to normal levels of 1–3 %. Infants are most susceptible to this effect because: 1) their drinking-water intake is relatively high in relation to their body weight, 2) fetal haemoglobin is more readily oxidized to methaemoglobin than the adult form, 3) stomach pH and gastro-intestinal conditions are favorable for reducing nitrate to nitrite, 4) they lack enzyme activity to reduce methaemoglobin back to haemoglobin. It is understood that most cases of infant methaemoglobinaemia due to high nitrate intake are associated with concomitant gastro-intestinal infections giving rise to an increased conversion of nitrate to the causal agent nitrite.

Another health risk possibly associated with high intake of nitrate in humans is the reaction of nitrite formed by reduction of nitrate with secondary amines to form carcinogenic N-nitroso compounds. Several N-nitroso compounds have been shown to be carcinogenic in animal species, but others such as the readily formed N-nitrosoproline are not carcinogenic in humans [12]. Epidemiological studies have presented conflicting results, showing both positive associations between gastric cancer and high nitrate intake as well as inverse or absence of associations. This brought the WHO to the conclusion that there was no firm evidence for a link between increased incidence of gastric cancer and high nitrate levels in drinking water but it was noted that the available data were inadequate [12, 13]. For other types of cancer, there are no data available to establish any association with nitrate or nitrite intake [12, 14]. Other important information related to the possible carcinogenic risk of high nitrate intake was provided by Kleinjans et. al. [15] showing that nitrate contamination of drinking water leading to an increased nitrate body load in exposed populations was not associated with chromosomal damage in peripheral lymphocytes.

In summary, there is insufficient epidemiological evidence for an association between nitrate intake and cancer, and therefore the current WHO guideline for nitrate in drinking water is solely established to prevent methaemoglobin formation. Epidemiological data support a guideline value of 50 mg/l for nitrate. Based on the information that also nitrite is present in some water supplies and the recognition of the causal role of nitrite in the formation of methaemoglobin, the WHO [2] also proposed a guideline value for nitrite. Based on the assumption that the relative potency of nitrite to cause this effect is 10 times higher than for nitrate (on a molar basis), a guideline value of 3 mg/l was derived. Because nitrate and nitrite might simultaneously occur in drinking water the following algorithm was derived:

$$\frac{C_{nitrate}}{GV_{nitrate}} + \frac{C_{nitrite}}{GV_{nitrite}} < 1$$

Where C = concentration nitrate or nitrite, and GV = guideline value for nitrate or nitrite.

6
Pesticides

Pesticides can reach ground- or surface water by leaching or runoff following normal agricultural use or by accidental (or deliberate) spills. In the latter case, sometimes large quantities may enter surface water leading to high concentrations. Although a wide variety of pesticides, herbicides, and fungicides have been detected, in general, levels in ground- and surface water are low and seldomly exceed 10 ppb [16]. It must be emphasized, however, that conventional drinking water treatment is poor at removing many of these pesticides because it was not specifically designed for this purpose.

Due to their nature, which is to kill or affect some form of life, pesticides are particularly considered to be a health hazard by the general public. In general, it is the public's perception that pesticides should not be present in drinking water and therefore often a precautionary principle is applied setting standards as low as reasonably achievable. By using a generally acceptable limit of determination for all pesticides, the EU has established limit values for individual pesticides of 0.1 µg/l and 0.5 µg/l for total pesticides [17]. The World Health Organization followed a different approach and derived health based guidelines for pesticides based on the available toxicological information. Guideline values for 33 pesticides usually ranged from 1 to 30 µg/l. Exceptions were at the higher end 2,4-DB with 90 µg/l and pyridate and dichlorprop with a guideline value of 100 µg/l, and at the lower end chlordane with 0.2 µg/l and heptachlor, aldrin and dieldrin with 0.03 µg/l. For a large number of these pesticides an allocation of the TDI of 1% was used for the derivation of the guideline value (in comparison with a default figure of 10%) to correct for the considerable exposure from other routes, in particular food. It was argued that these guidelines might not be conservative enough since they do not take into account the potential toxicity of degradation products and it was recommended that such degradation products should be taken into consideration [2]. This issue will be particularly considered during the current revision of the guidelines [5]. It was recognized, however, that in most cases inadequate data were available on the identity, the presence, and the biological activity of these degradation products.

7
Disinfectants and Disinfection By-Products

The most important step in the production of safe drinking water is without doubt disinfection. Waterborne diseases and water-related death are still a considerable problem in developing countries. Diarrheal disease is one of the most important health problems related to waterborne pathogens but amoebiasis and schistosomiasis also contribute considerably to the morbidity and mortality rate in developing countries [18]. Destruction of pathogenic microorganisms is an essential tool in fighting these diseases. Disinfection of drinking water involves the use of very reactive chemicals such as chlorine which has been the most important disinfectant for decades or ozone, which is now being increasingly used in drinking water treatment. These compounds are not only very powerful biocides but, due to their reactivity, they also react with organic micropollutants in drinking water and thus give rise to the so-called disinfection by-products (DBPs). There is a wide range of by-products caused by the reaction of chlorine with humic substances in the water. The most common DBPs are bromoform, chloroform, bromodichloromethane, and dibromochlo-

Table 2. WHO Guidelines for carcinogenic disinfectant by-products (2)

bromate	25 µg/l	for excess cancer risk of 7×10^{-5}
bromodichloromethane	60 µg/l	for excess cancer risk of 10^{-5}
chloroform	200 µg/l	for excess cancer risk of 10^{-5}

romethane. The by-products of the use of ozone are less well characterized, but one of the well know DBPs is bromate.

A number of these unwanted DBPs cause major health concern because of their carcinogenic properties. The World Health Organization has set guidelines for these compounds based on an excess cancer risk of 1 in a population of 100 000 (Table 2), under the assumption, as shown previously, that the use of mathematical low-dose risk extrapolation in general provides an overestimation of the actual risk. In addition, it should be noted that several epidemiological studies have been carried out to investigate the possible carcinogenic properties of chlorinated drinking water, but the International Agency for the Research on Cancer considered that the degree of evidence for an association between chlorination and the occurrence of cancer from these studies is inadequate [19]. The World Health Organization emphasized that when a choice has to be made between meeting guidelines for disinfectants and disinfectant by-products on one hand and micribiological guidelines on the other, the microbiological quality must take preference. Consequently it stated that "efficient disinfection must never be compromised" [2]. From a practical point, however, one should realize that the level of harmful disinfectant by-products can be substantially reduced by the removal of organic substances from drinking water prior to the use of disinfectants.

References

1. WHO (1992) Our planet, our health, Report of the WHO Commission on Health and Environment, Geneva, Switzerland
2. WHO (1993) Guidelines for Drinking-Water Quality, 2nd edn: Recommendations, vol 1, Geneva, Switzerland
3. Van Dijk-Looiijaard A (1995) In: Hrubec J. (ed), Quality and treatment of drinking water, The Handbook of Environmental Chemistry, vol 5B, Springer, Berlin Heidelberg New York
4. WHO (1984) Guidelines for Drinking-Water Quality, Recommendations, vol 1, Geneva, Switzerland
5. WHO (1996) Rolling Revision of the WHO Guidelines for Drinking-water Quality, Report WHO/EOS/ECEH/96.1, Geneva, Switzerland
6. OECD (1987) Guidelines for the Testing of Chemicals, OECD, Paris, France
7. WHO (1991) 38th report of the Joint FAO/WHO Expert Committee on Food Additives, WHO Technical report Series 815, Geneva, Switzerland
8. International Programme on Chemical Safety (1992) Report on IPCS discussions on deriving guidance values for health-based exposure limits, PCS/92.6, Geneva, Switzerland
9. Cotruvo JA (1988) Regulatory Toxicol and Pharmacol 8:288
10. Crump K, Allen B, Shipp A (1989) Health Phys. 57 (Suppl. 1):387
11. WHO (1987) Report of the 30th Meeting of the Joint FAO/WHO Expert Committee on Food Additives, WHO Food Additives Series 21, Geneva, Switzerland

12. WHO (1996) Report of the 44th Meeting of the Joint FAO/WHO Expert Committee on Food Additives, WHO Food Additive Series 35, Geneva, Switzerland
13. WHO (1985) Report on a WHO meeting, Environmental Health Series No.1, WHO Regional Office, Copenhagen, Denmark
14. Gangolli SD, van den Brandt PA, Feron VJ, Janzowsky C, Koeman JH, Speijers GJA, Spiegelhalder B, Walker R, Wisnok JS (1994) Eur J Pharmcol, Environ Toxicol Pharmacol Section, 292:1
15. Kleinjans JCS, Albering HJ, Marx A, van Maanen JMS, van Agen B, ten Hoor F, Swaen GMH, Mertens PLJM (1991) Environ Health Perspect 94:189
16. Fawell JK (1991) BCPC Monograph No. 47, p 205
17. Council of the European Communities (1980) Directive relating to the quality of water intended for human consumption, 80/778/EEC, Official Journal L229, Brussels, Belgium
18. Galal-Gorchev H (1993) In: Craun GF (ed) Safety of water disinfection: Balancing chemical and microbiological risks, ILSI Press Washington, DC, p 463
19. International Agency for the Research on Cancer (1991) IARC Monograph on the evaluation of carcinogenic risks to humans, No. 52, Lyon, France

Toxicity Tests for Assessing Drinking Water Quality

J.K. Fawell and H. Horth

WRc plc, Henley Road, Medmenham, Marlow, Bucks, SL7 2HD, UK
E-mail: fawell@wrcplc.co.uk

Drinking water contains a range of natural and anthropogenic contaminants which vary between water sources and with time. Toxicity tests based on bacteria, cells in tissue culture, plants and even vertebrates in vivo have been used to screen drinking water or concentrates from drinking water. These tests use a variety of endpoints from basal cytoxicity to effects on DNA. Although they cannot be used to directly estimate risk to man, they are valuable tools and have been used successfully, particularly in research on drinking water treatment.

Keywords: water quality, toxicity, cytotoxicity, genotoxicity, bioassays.

Contents

The Handbook of Environmental Chemistry Vol. 5 Part C
Quality and Treatment of Drinking Water II (ed. by J. Hrubec)
© Springer-Verlag Berlin Heidelberg 1998

1
Introduction

Drinking water contains a variety of inorganic and organic contaminants depending on its source. The presence of a range of both natural and anthropogenic contaminants, which varies both qualitatively and quantitatively, presents a number of difficulties for monitoring drinking water quality. As our knowledge of the constituents of drinking water has increased, the idea of a simple screening test which can give an indication of the likely risks to health of the sum of contaminants in a particular water has also become more attractive.

There have been attempts to use a range of test systems with varying biological endpoints and with varying degrees of success. The systems vary from simple cells such as bacteria to aquatic vertebrates and from simple endpoints such as cell death to effects on the genetic material of the cell. Unfortunately many of the problems which face analytical chemists such as method specificity and sensitivity also apply to biological tests. However, just as analytical chemistry is used successfully when applied properly, so too are toxicity tests, even though the original aim may not be fully realised.

2
Problem Areas

As mentioned above, there are several broad areas of difficulty which must be considered when assessing the value of toxicity tests for examining drinking water. Much will depend on the objectives of carrying out the test and this is of key importance. For example, the requirements for a screening test in research will be very different from those if the test is to be used operationally. Similarly the requirements for screening for a specific compound or group of compounds will be different from a broad screen for potential toxicity. The particular considerations are discussed below.

2.1
Long-Term Exposure

For most contaminants present in drinking water the major concerns relate to long-term, even lifetime, exposure. However, most of the toxicity tests such as cytotoxicity or bacterial toxicity tests relate to short-term exposure with a measure of basal cell toxicity. Only one long-term concern can be appropriately addressed and that is *potential* genotoxic carcinogenicity which is indicated by mutagenic activity in a variety of organisms. This has been used with some success in a research context.

It may be possible to use screening tests to act as a fail-safe for detecting chemical contaminants, for example those which have been spilt to surface waters in high concentration but which are not known to be present. However, this requires that the sensitivity of the test be adequate, that samples are taken at the right time and that the assay can give results sufficiently rapidly to take action to prevent the pollutant entering supply.

2.2
Sensitivity

For most substances the sensitivity of short-term toxicity assays is insufficient for screening drinking water. The concentrations which would be of concern are usually well below those causing clearly discernible effects in biological test systems if a margin of safety is desired. This is demonstrated by examining drinking water standards and the guidelines for potentially toxic substances. For example, the proposed WHO guideline values, which take into account long-term toxicity, for dichloroacetic acid (DCA) and trichloroacetic acid (TCA) respectively are 50 and 100 μg l^{-1} [1]. However, in cytotoxicity assays the EC_{50} values for these chemicals were in excess of 1 mg ml^{-1} [2]. This reflects the difference between long and short-term exposure since the rat oral LD_{50}s are also very high.

2.3
Specificity

In contrast to analytical techniques, the requirement for toxicity screening tests is that they are not specific to individual, or groups, of substances. The specificity of a toxin may be because it is accumulated in a particular organ or affects a very specific biochemical pathway. Unless this can be replicated in an in vitro system, then the toxin will not be detected.

2.4
Risk Assessment

This is the most difficult problem facing the use of toxicity tests for assessing drinking water quality. In vitro systems which are potentially very different from humans, i.e. non-mammalian, cannot be readily used for risk assessment. This is particularly true when one is dealing with complex mixtures of unknown compounds and when the data relate to short-term exposure although the interest is in long-term exposure. In carrying out a risk assessment on a toxic substance, a wide variety of toxicological data must be considered. This includes metabolism, long and short-term toxicity and reproductive toxicity. When this information is used to derive a guideline value or standard, then the uncertainties in the extrapolation and the data are reflected in the guideline value so that there is a significant margin of safety for all consumers.

Even a positive finding in a mutagenicity assay does not either confirm that there is a hazard to man or give any quantitative indication of the potential risk.

3
Cytotoxicity Tests

There has been considerable development of in vitro toxicity tests based on mammalian cells in tissue culture resulting in a variety of test systems. These assays are most frequently based on continuous cell lines but only a few, such as the Hep G2 cell line, possess significant capability for metabolising foreign compounds. However, these do not yet appear to have been used to assess the toxicity of water samples. In general the sensitivity of the cell line appears to be less important than the ease of culture and the applicability of the end point. The broad application of these assays in the water industry has been reviewed by Hunt et al. [3].

From 1964 to 1973 the Metropolitan Water Board in London used an assessment of morphological changes in Vero cells, which were adapted from virus culture, to assess the cytotoxicity of a range of treated waters [4]. Only one of several hundred drinking water samples gave a positive result which was found to be due to a chemical discharge from a factory [5].

At the same time similar studies were undertaken at the Health Laboratory of the City of Paris [6]. Concentrated chloroform extracts of waters were studied for their ability to produce morphological changes in the cells with the result that surface water appeared to be of generally greater cytotoxicity than groundwaters [7].

Inhibition of cell growth has also been used as an endpoint for cytotoxicity, although few unconcentrated drinking water samples have been assessed. Most of the studies published used concentrated extracts and most of the effects observed were relatively small [8–10].

Cloning efficiency, or the ability of individual cells to form colonies, is a sensitive indicator of cytotoxicity. This has been used with Buffalo green monkey kidney cells and mouse lymphoma cells to assess the suitability of wastewaters for reuse as drinking water in South Africa and, in particular, to examine the removal of cytotoxicity by various treatment stages [11–12]. This technique was also used in the USA to assess the suitability of different concentration methods, with reverse osmosis concentrates being up to 20 times more toxic than eluates from XAD-2 resin [13].

Other endpoints have included oxygen uptake by cells in culture for examining wastewater [14] and inhibition of RNA and DNA synthesis. Although most studies have used river water [15–16] one group have used the method to study the effects of different kinds of drinking water treatment on cytotoxicity. In this study, granular activated carbon (GAC) was ineffective, ozone and chlorine dioxide reduced activity, ozone and GAC abolished activity while chlorine after GAC increased activity [17].

Cytotoxicity assays are used to test materials which are to come into contact with drinking water, usually by testing the leachate from such materials [18–20].

Hunt et al. [2] carried out a study to compare the effectiveness of four tests for application to water samples. They looked at 13 compounds in each test and compared the sensitivity. They also examined six water samples, two river waters, three sewage effluents and one industrial effluent in two of the assays, cloning efficiency and inhibition of cell growth by measuring vital dye (neutral red) uptake. The sensitivity of the assays was low in relation to the possible concentrations of interest in drinking water, for example the provisional WHO guideline values for dichloroacetic acid and trichloroacetic acid are 50 and 100 μg l^{-1} respectively, while the most sensitive cytotoxicity assay would only detect these compounds at concentrations in excess of 1 mg l⁻¹. The difference reflects the concern for possible long-term effects from substances of low acute toxicity, when setting standards or guidelines. The most sensitive endpoint for assessment of the cytotoxicity of the water samples was cloning efficiency after an exposure period of 8–10 days. This demonstrates that the extended exposure period is a key factor in improving sensitivity with regard to testing water samples.

Beresford and Good [21] attempted to develop a more sensitive and rapid assay for use in screening for waterborne contaminants. They recommended the use of the neutral red (vital dye) uptake method and the MTT test which is based on the metabolic reduction of 3-(4,5-dimethylthiazol-2-yl)-2-5-diphenyl tetrazolium bromide to a blue formazan product by succinate dehydrogenase in the mitochondrial membrane of viable cells. They also measured the ATP content of cells using a test kit based on the luciferin/luciferase assay [22]. They concluded that, by using the test water to dilute concentrated culture medium, the sensitivity could be maximised but that pre-concentration procedures would probably be required.

In the United States, embryonic cultures were used to assess the cytotoxicity and biochemical effects of particulate matter isolated from six drinking water samples. The aim of the study was to compare the cytotoxicity of various particulate samples from water with commercial chrysolite asbestos [23].

More recently a group in Japan have used cytotoxicity tests, in combination with mutagenicity tests, on concentrated extracts to examine raw, waste and treated drinking waters, although no attempt was made to identify the components of the extracts [24–25].

4
Bacterial Toxicity Assays

A number of assays have been developed for toxicity testing based on bacteria [26–28], however, the most widely used is the Microtox testing system [29–30]. This system is commercially available [31] and uses the marine bacterium *Photobacterium phosphorous* which emits light as a natural by-product of respiration. The reduction in light emitted is proportional to the inhibition of metabolism and this is used as a measure of toxicity. There are many factors which are important in bacterial assays including the permeability of the cells to various potential toxicants [32]. Most of the bacterial toxicity assays originate from the determination of the toxicity of various substances to sewage organisms [33] but they have also been developed for use in testing for waterborne toxicants under other circumstances. They can be used effectively and rapidly but suffer from the same shortcomings as cytotoxicity tests. Nevertheless they are proving to be very useful in a number of areas although rather surprisingly they have not been widely applied to drinking water. In particular they are more convenient and simple to use in comparison with cells in tissue culture and technically much less demanding. It must also be said that they are probably no less relevant to whole animals than simple tissue culture systems when looking at acute toxicity.

Slabbert [27] developed a toxicity assay based on oxygen uptake by *Pseudomonas putida* intended for a range of uses with aquatic samples including contaminated drinking water. In his comments, the author remarks that "being a very rapid test, and carried out under defined laboratory conditions, the test cannot be used to predict environmental toxicity." The important word here is predict.

Chou and Que Hee [34] used the Microtox test to determine EC50 values for by-products of ozonation found in drinking water. The substances concerned could not be detected at the concentrations found in drinking water. However, it must be emphasised that these concentrations may be well below those which would be of concern to human health.

The position with regard to cytoxicity tests also applies to bacterial tests. They do have an advantage with regard to simplicity of use but they will not detect effects directly on the cell membrane with the efficiency of tests using mammalian cells. However, as long as the constraints are understood they can be a very useful tool in drinking water research and development.

5
Biochemical Assays

Blondin et al. [35] proposed a test using isolated mammalian mitochondria which were disrupted to release sub-mitochondrial particles. The bio-assay was based on an energy-coupled reverse electron transfer in which toxicity is measured by inhibition of the rate of NAD^+ reduction. The authors recommended the test should be used as a means of replacing fish tests, but recommendations were also made with regard to drinking water.

6
Assays for Genotoxicity

Changes in the DNA of a cell transmitted to daughter cells when the parent cell divides, can give rise to significant changes in the physiological response of the cell. If the cell in which the mutation occurs is a germ cell then this could lead to heritable diseases being passed on to offspring. If the cell in which the mutation occurs is a somatic cell then this could trigger a long series of events leading to the development of cancer.

Assays in genetically manipulated strains of *Salmonella typhimurium* were developed by Bruce Ames and co-workers [36] to study mutagenicity, and these assays formed the basis of a huge explosion in studies to screen a wide variety of substances for their ability to cause mutations. Further assays were developed using other bacteria and yeast and still more tests were devised to examine chromosome damage and mutations in mammalian cells in tissue culture, mutations in the fruit fly, *Drosophila*, and in whole animals.

These assays are now well established for use with chemicals but are less well established for use with mixtures. This review and others [37] concentrate on those assays which have been used to screen drinking water for genotoxic activity.

7
Mutagenicity of Drinking Water

Following the identification of a number of anthropogenic substances in drinking water both arising from the use of surface waters containing effluents and as a consequence of chlorination, attempts were made to use assays for mutagenicity as a tool to study drinking water contamination [38–40]. Initially attempts were made to test unconcentrated samples using a very sensitive assay, the liquid based fluctuation assay [41–42], but most studies made use of some form of concentration procedure. Early studies used bacterial assays, particularly based on *S. typhimurium* strains TA 98 and TA 100 with and without the presence of rat liver S-9 mix to simulate mammalian metabolism, which may be required to change some mutagenic substances to the active form. Later, other assays including mammalian cells in vitro, *Drosophila*, and whole animals, using either mammals or amphibians were used.

7.1
Bacterial Assays

Initial studies using unconcentrated drinking water samples in the fluctuation assay mentioned above gave only very weak positive results in the absence of S-9 mix [42–43]. This meant that the validity of the data was uncertain and, in addition, it was demonstrated that nutrients naturally present in the water enabled the bacteria to utilise the available histidine more efficiently than the controls. Since the assay was based on reversion of the bacteria from a form unable to utilise histidine to a wild form which could grow without added histidine, this feature could give rise to false positive results.

Subsequent studies were carried out using some form of concentration procedure. A number of possibilities such as freeze drying [44] and liquid/liquid extraction [45] were used, but most studies relied on extraction of the contaminants from water by adsorption to various macroreticular (XAD) resins with subsequent solvent elution. A variety of XAD resins, including 2, 4 and 8 and mixtures of resins were tried by different workers along with a range of eluting solvents, but results were broadly similar. Each system had advantages and disadvantages, including toxicity of extracts to the bacteria. Extraction and concentration procedures were reviewed in detail by Wilcox et al. [46]. There was some uncertainty with all the techniques used, including XAD, since all, to a different extent, are selective in the range of organic substances which they will extract and concentrate. There were also practical difficulties in preparing suitable experimental blanks and controls, since it is difficult to account for any reactions taking place in the concentration procedure which could change the nature of the contaminants and give rise to artifacts. In drinking water samples one of the obvious problems is residual chlorine, which could react with the resins, but the use of dechlorinating agents such as sodium bisulphite were also shown to affect mutagenic substances present in the samples [47–48].

7.2
Examples of the Use of Bacterial Mutagenicity Assays

Many groups in different parts of the world have carried out research on drinking water using such techniques [49–55]. In general there is a considerable level of consistency in the data, particularly for chlorinated drinking water.

Extracts of chlorinated drinking water were shown to be positive in S. typhimurium strain TA 100 and TA 98 in the absence of rat liver S-9. In the presence of S-9, activity was reduced or removed. Mutagenic activity was particularly strong in TA100. The level of activity was similar with extracts from both freeze dried and XAD adsorbed samples [56], which supports the contention that XAD, despite its selectivity, was able to concentrate many of the genotoxic components of the mixture.

Initial studies made no adjustment to the pH of the sample (about 7.0) in the adsorption stage but subsequent studies showed that acidification of the column effluent (i.e. the water sample) to pH 2.0 and passage through a second column, gave extracts of similar or greater mutagenic activity [57–58]. This seemed to indicate that a complex mixture of mutagens was present in chlorinated drinking water.

Studies with indirect re-use of wastewater (i.e. discharges of treated sewage effluent to a river used as a source for drinking water) showed little activity in the raw water [55], but a substantial increase in mutagenicity following chlorination. Chorinated upland waters containing high levels of naturally occurring humic substances also showed high levels of activity [59] and chlorinated humic and fulvic acid solutions showed the same pattern of activity [54, 55, 60]. Fractionation techniques with mutagenicity assays to isolate active fractions were used to help identify some of the mutagenic substances [61].

However, it was studies on chlorinated pulp mill effluent which led to the identification of a low molecular weight polar substance, 3-chloro-4-(dichloromethyl)-5-hydroxy-2(5H)-furanone (MX) having very high mutagenic activity [62]. This substance was subsequently also identified in drinking water [63]. MX is highly mutagenic in S. typhimurium TA 100 in the absence of rat liver S-9 fraction and, at a few nanograms per litre, contributes between 30 and 60% of the XAD-extractable mutagenicity from chlorinated drinking water [56].

A number of other studies have been carried out using bacterial or yeast assays for mutagenicity which are not directly associated with research into mutagens in chlorinated drinking waters. Kool et al. [64] used the Ames Salmonella S9 assay to study mutagens in groundwater arising from the mobilisation and leaching of organic mutagens in soil. Basu et al. [65] used bacterial mutagenicity in S. typhimurium TA 1535, TA 1537, TA 1538, TA 98 and TA 100 to study the leaching of polycyclic aromatic hydrocarbons from coal-tar lined distribution pipes. Some workers studied alternative disinfectants to chlorine [66–69], while others used bacterial mutagenicity to study the impact of pollutants in raw water [70–71] and the ameliorating effect of humic acid on raw water mutagens [72]. Mutagenicity in bacteria has also been used to evaluate treatment processes for drinking water as an additional parameter alongside traditional analytical chemistry [73–75].

7.3
Assays in Cells in Tissue Culture

Cells in culture provide a higher level of test for mutagenicity since the cell membrane is intact and largely unaltered, unlike many of the strains of bacteria used for mutagenicity testing. It is also possible to assess the ability of genotoxic substances to impact at a supra-molecular level by measuring chromosome damage. However, most cell lines still require the addition of S9 or some other metabolic system to detect substances which require metabolic activation to the active mutagen.

There are a number of in vitro assays available for assessing chromosome damage which include cytogenetics in Chinese hamster ovary cells (CHO cells) and human lymphocytes. In the past, some workers have favoured measurement of sister chromatid exchange (SCE) but although this appears to be very sensitive, the significance of SCE as a genotoxic endpoint remains uncertain. Tests using higher cells in culture are much more technically difficult to carry out and measurement of the endpoints, particularly chromosome abberations, requires trained personnel. Their use in examining water samples has therefore been much more limited and has primarily focussed on confirming the data obtained using bacterial assays.

Studies at the Water Research Centre [56, 76, 77] were carried out using CHO cells and XAD resin extracts of laboratory chlorinated samples from drinking water sources. Initial studies with extracts adsorbed at pH 7.0 indicated low cytogenetic activity. However treatment of the cells in serum-free medium for three hours before serum was added, resulted in a much higher level of activity. This was because the mutagens appeared to be reacting with the serum. Subsequent studies showed that the potent bacterial mutagen MX did bind to serum proteins in vitro [78]. The findings were confirmed on samples from lowland river sources and upland reservoirs, with both final drinking water samples and laboratory chlorinated samples of raw water examined. These confirmatory studies included experiments in which a series of different concentrations of exogenous protein were added to an extract from a chlorinated drinking water sample. There was a clear inverse dose response with increasing protein concentration [56]. The production of mutagens from the reaction of chlorine with humic acid solutions was also confirmed in CHO cells.

Studies in human lymphocyte cultures were carried out to confirm the data from CHO cells. In this case whole blood cultures gave no response but cultures of separated lymphocytes also gave positive results with extracts of chlorinated drinking water [76]. Varga [79] also used cytogenetics and SCE in peripheral human lymphocytes in cultures to examine concentrated extracts of drinking water, but only weak activity was seen with SCE induction.

The detection of point mutations in mammalian cells in culture is possible, but when a substance or a mixture of substances is also clastogenic, there can be difficulties in differentiating the two. However, there are assays with specific endpoints which do enable the two mechanisms to be differentiated.

Studies with concentrated extracts of drinking water gave positive results in the mouse lymphoma mutagenesis assay [80–81]. Other workers compared the activity of drinking water concentrates, extracted at pH 7.9, in the mouse lymphoma assay, in Chinese Hamster V79/4 cells and CHO cells using point mutation endpoints [77]. The extracts showed marginal activity in the three assays but a high frequency of structural chromosome aberrations were observed in parallel experiments with the same cells.

7.4
Assays for Genotoxicity in Vivo

There are several assays for genotoxicity which utilise intact animals ranging from insects through aquatic invertebrates and vertebrates to mammals. A number of these have been used to test both unconcentrated and concentrated extracts of drinking water.

Concentrated XAD extracts of chlorinated drinking water (at pH 7.0) were studied for their ability to cause sex-linked recessive lethal mutations in *Drosophila*. These studies optimised exposure so that toxicity was apparent, indicating that an appropriate level of exposure was achieved, but there was no increase in mutations [82].

Tests using aquatic invertebrates have been used to examine mutagenicity of wastewaters [83] but none of these appear to have been used to examine drinking water. However, the induction of micronuclei in blood cells of the larvae of the newt, *Pleurodeles watl*, have been used to examine chlorinated and ozonated drinking water [84–85]. Animals are exposed for longer periods in this assay since the larvae can be maintained in the test environment. Positive results were obtained with chlorinated drinking water and from samples of water treated with low doses of ozone.

A number of studies have examined the production of micronuclei or chromosome abberations in mouse bone marrow following dosing with concentrated extracts of chlorinated drinking water. When mice were dosed with XAD extracts of water adsorbed at pH 7.0, equivalent to average human drinking water consumption over one, three and ten years, there was no indication of cytogenetic damage [76]. Similar negative results were reported from studies on the effects of dosing concentrated extracts of water treated with chlorine and alternative disinfectants on micronuclei of mouse bone marrow cells [66].

Subsequent studies examined nuclear anomalies from apoptosis, karyorrhexis and pyknosis, in the gastro-intestinal tract and urinary bladder of mice following dosing with concentrated extracts equivalent to 100 l per mouse [56]. These results were negative.

8
Assays in Plants

Since plants can be grown in nutrient rich water streams and there are various assays for toxicity and genotoxicity which utilise plants, particularly of the onion family, their potential for screening water samples is clear. Fiskesjö 1993 [86] published a toxicity test based on measuring mean root growth in *Allium cepa* L. after exposure for 2–3 days. Heartlein et al. 1981 [87] utilised a maize reverse-mutation test to examine raw and treated waters from an agricultural district but were unable to detect any genotoxicity. Other workers used the Allium anaphase-telophase test for screening wastewaters [88] for genotoxicity but did not use this technique for drinking water.

A more sophisticated system for the estimation of toxicity in water has been proposed by Ludiyanskiy and Pasichuy 1992 [89]. This uses both aquatic vascular plants and algae and is based on photosynthesis and respiration intensity.

9
Discussion

There are a wide range of tests which have been used to assess a range of different toxic endpoints. However, none can be used to assess directly the risks to man from individual substances or mixtures of substances in drinking water. One of the difficulties highlighted in the problem areas section above, was that of detecting toxicity arising as a consequence of long-term exposure in man, with tests relying on short-term exposure. In addition the test systems are mostly very simple and so extrapolation to man requires a number of large assumptions. It is important to remember that in order to properly assess the risks to man from contaminants in drinking water, it is necessary to identify the contaminants and consider a wide range of biological activity.

Genotoxicity is considered to be a marker for hazards arising from long-term exposure, i.e. carcinogenicity and genetic damage. However, such tests give only an indication of the potential for genotoxicity. A positive finding in an in vitro assay requires confirmation in a whole animal test, preferably a mammal, before it is considered to indicate a potential hazard for man. However this gives no indication of the risks associated with exposure to that hazard. MX for example is an extremely potent mutagen in bacteria, but has little activity in vivo.

One of the difficulties encountered is that, in general, the concentration of contaminants in drinking water is low and delivering a sufficient dose to cause toxicity requires either long-term exposure or the use of concentrated extracts. The former is not practicable, while the latter requires great care to minimise false positives and negatives. It is particularly important that appropriate controls are used, including methodological blanks. In assessing genotoxicity in complex mixtures there is an additional uncertainty since toxic substances (which are not genotoxic) may prevent the use of sufficiently high doses in test systems to detect the genotoxic components of the mixture.

Care must also be exercised in the interpretation of results. Where possible, the normal principles of toxicology should be applied and a dose response should be achieved to have confidence in a positive result. In the case of genotoxicity assays then a sufficiently high response above background would also be required. In the case of the Ames test with various strains of S. *typhimurium* this would be a minimum doubling of the number of revertant colonies. However, purely statistical differences could be used for comparative purposes, for example, in studying different treatment processes.

As long as toxicity tests are used with care and within their limitations then they provide an important tool in studying contamination of drinking water. At the beginning of the 1980s following the development of a range of toxicity tests and assays, there was a subsequent explosion in the number of groups using them to assess water quality. This has significantly declined, at least with regard to published studies. However, such tests still have an important function in assessing drinking water quality although they are more appropriate in an investigative role rather than for routine monitoring.

References

1. World Health Organisation (1993) Guidelines for Drinking Water Quality. 2nd edn. Volume 1, Recommendations. WHO, Geneva
2. Hunt SM, Chrzanowska C, Barnett CR, Brand HN, Fawell JK (1987) A comparision of in vitro cytotoxicity assays and their application to water samples. ATLA, Alternatives to Laboratory Animals 15:20
3. Hunt SM, Chrzanowska C, Barnett CR, Fawell JK (1986) The application of in vitro cytotoxicity assays to the water industry. ATLA, Alternatives to Laboratory Animals 14:64
4. Metropolitan Water Board (1964) 41st report on the results of the bacteriological, chemical and biological examination of the London waters for the years 1963–1964
5. Metropolitan Water Board (1966) Forty-second report on the results of the bacteriological, chemical and biological examintion of the London waters for the years 1965–1966
6. Coin L (1963) Quelques resultats en matière de virologie et le problème de la cytotoxicité. La Technique de l'Eau 17(200):27
7. Coin L, Hannoun C, Trimoreau JC (1969) The problem of the cytotoxicity of water. In:Proceedings of the 4th International Conference on Water Pollution Research, Prague, p 95
8. Mochida K (1986) Aquatic toxicity evaluated using human and monkey cell culture assays. Bulletin of Environmental Contamination and Toxicology 18:683
9. Cody TE, Elia VJ, Clark CS, Christian RT (1979) Integrated use of bioassays and chemical analyses to evaluate the quality of reuse water. In: Proceedings of Water Reuse Symposium. Water Reuse – From Research to Application, p 2230. AWWA Research Foundation, Denver, USA
10. Maruoka S (1978) Estimation of toxicity using cultured mammalian cells of the organic pollutants recovered from Lake Biwa. Water Research 12:371
11. Kfir R, Prozesky OW (1981) Detection of toxic substances in water by means of a mammalian cell culture technique. Water Research 15:553
12. Kfir R, Prozesky OW (1982) Removal of toxicants during direct and indirect reuse of wastewater evaluated by means of a mammalian cell culture technique. Water Research 16:823
13. Loper JC, Lang DR, Scheny RS, Richmond BB, Gallagher PM, Smith CC (1978) Residue organic mixtures from drinking water show in vitro mutagenicity and transforming activity. Journal Toxicology Environmental Health 4:919
14. Slabbert JL, Steyn PL, Bateman BW, Kfir R (1984) Rapid detection of toxicity in water using the oxygen uptake rate of mammalian cells as a sensor. Water South Africa 10:1
15. Van Doren SR, Hall MS, Frazier LB, Leach FR (1984) A rapid cell culture assay of water quality. Bulletin of Environmental Contamination and Toxicology 32:220
16. Fauris C, Danglot C, Vilagines R (1985a) Rapidity of RNA synthesis in human cells. A highly sensitive parameter for water cytotoxicity evaluation. Water Research 19:677
17. Fauris C, Danglot C, Montiel A, Vilagines R (1985b) Toxicology evaluation of water treatments. Environmental Toxicology Letters 6:279
18. Ashworth J, Colbourne JS (1981) The effect of non-metallic materials employed in water supply distribution and plumbing systems, and consumer fittings upon potable water quality. International Conference Brighton, CEP Consultants, Edinburgh. Water Industry 81:130

19. United Kingdom Water Fittings Byelaws Scheme (1985) Requirements for the testing of non-metallic materials for use in contact with potable water. No 5-01-02, Issue 3
20. British Standard BS6920 (1990) British Standards Institution, London, UK
21. Beresford DJ, Good S (1987) Development of cytotoxicity tests for assessment of the toxicity of water samples taken from the environment. Molecular Toxicology 1: 419
22. Lundin A, Rickardsson A, Thore A (1976) Continuous monitoring of ATP converting reactions by purified firefly luciferase. Analytical Biochemistry 75:611
23. Hart RW, Fertel R, Newman HAI and Blakeslee JR (1985) Effects of selected waterborne particulates on cellular and molecular parameters. Government Reports, Announcements and Index (GRA&I) 5
24. Utsumi H, Kiyoshige K, Mitade C, Han KS, Hakoda M, Manabe H and Hamada A (1992a) Mutagenicity and cytotoxicity of tap and raw water and their potential risk. Water Science and Technology 26:247
25. Utsumi H, Hakoda M, Kiyoshige K, Manabe H, Mitade C, Muryama J, Han KS and Hamada A (1992b) Cytotoxicity and mutagenicity of micropollutants in drinking water. Water Science and Technology 25:325
26. Reinhartz A, Lampert I, Herzberg M and Fish F (1987) A new, short-term, sensitive, bacterial assay kit for the detection of toxicants. Toxicity Assessment: An International Quarterly 2:193
27. Slabbert JL (1987) Technical methods section – a toxicity assay based on oxygen uptake by *Pseudomonas putida*. Toxicity Assessment: An International Quarterly 2:229
28. Xu H and Dutka BJ (1987) ATP-TOX system – a new, rapid, sensitive bacterial toxicity screening system based on the determination of ATP. Toxicity Assessment: An International Quarterly 2:149
29. Bulich AA (1979) Use of luminescent bacteria for determining toxicity in aquatic environments. In: Markings LL and Kimerle RA (eds) Aquatic Toxicology, ASTM 667 American Society for Testing and Materials, p 98
30. Bulich AA (1986) Bioluminscence assays. In: Bitton G and Dutka BJ (eds) Toxicity Testing Using Micro-organisms. CRC Press, Boca Raton, Florida, p 57
31. Microbics Corporation (1982) Microtox system operating manual, Microbics Corporation, Carlsbad, California
32. Bitton G, Dutton RJ and Koopman B (1988) Cell permeability to toxicants: an important parameter in toxicity tests using bacteria. CRC Critical Reviews in Environmental Control 18:177
33. Painter HA (1993) A review of tests for inhibition of bacteria (especially those agreed internationally). In: Richardson M (ed) Ecotoxicology Monitoring VCH, p 17
34. Chou CC and Que Hee SS (1992) Microtox EC_{50} values for drinking water by-products produced by ozonolysis. Ecotoxicology Environmental Safety 23:355
35. Blondin GA, Knobeloch LM, Read HW and Harkin JM (1987) Mammalian mitochondria as in vitro monitors of water quality. Bulletin of Environmental Contamination and Toxicology 38:467
36. Ames BN, McCann J and Yamasaki E (1975) Methods for detecting carcinogens and mutagens with the Salmonella/mammalian microsome mutagenicity test. Mutation Research 31:347
37. Meier JR and Daniel FB (1990) The role of short-term tests in evaluating health effects associated with drinking water. Journal of American Water Works Association 82:48
38. Loper JC, Lang DR, Smith CC, Schoeny RS, Kopfler FC and Tardiff RG (1978) In vitro mutagenesis and carcinogenesis testing of residual organics in drinking water. In: Hutzinger O, Van Lelyveld LH and Zoeteman BCJ (eds) Aquatic Pollutants. Transformation and Biological Effects. Pergamon, p 405
39. Loper JC (1980) Mutagenic effects of organic compounds in drinking water. Mutation Research 76:241
40. Forster R and Wilson I (1981) The application of mutagenicity testing to drinking water. Journal of the Institute of Water Engineers and Scientists 35:259

41. Green MHL, Muriel WJ and Bridges BA (1976) Use of a simplified bacterial fluctuation test to detect low levels of mutagens. Mutation Research 33 : 33

42. Forster R, Green MHL, Gwilliam RD, Priestley A and Bridges BA (1983) Use of the fluctuation test to detect mutagenic activity in unconcentrated samples of drinking waters in the United Kingdom. In: Jolley RL et al. (ed) Water Chlorination: Environmental Impact and Health Effects. Ann Arbor Science, Michigan

43. Harrington TR, Nestmann ER and Kowbel DJ (1983) Suitability of the modified fluctuation assay for evaluating the mutagenicity of unconcentrated drinking water. Mutation Research 120:97

44. Forster R (1984) Mutagenicity testing of drinking water using freeze-dried extracts. In: Microbiological Methods for Environmental Biotechnology. Society for Applied Bacteriology

45. Grabow WOK, Burger JS and Hilner CA (1981a) Comparison of liquid-liquid extraction and resin adsorption for concentrating mutagens in Ames Salmonella/microsome assays on water. Bulletin of Environmental Contamination and Toxicology 27:442

46. Wilcox P, van Hoof F and van der Gaag MA (1986) Isolation and characterisation of mutagens from drinking water. In: Leonard A and Kirsch-Volders M (eds) Proceedings of XVth Annual Meeting of European Environmental Mutagen Society. Brussels, p 92

47. Cheh AM, Skochdopole J, Koski PM and Cole L (1980) Nonvolatile mutagens in drinking water: production by chlorination and destruction by sulfite. Science 20:90

48. Wilcox P and Denny S (1984) Effect of dechlorinating agents on the mutagenic activity of chlorinated water samples. In: Jolley RL et al. (eds) Water Chlorination, Chemistry, Environmental Impact and Health Effects Volume 5. Lewis, Michigan, p 1341

49. Nestmann ER, Le Bel GL, Williams DT and Kowbel DJ (1979) Mutagenicity of organic extracts from Canadian drinking water in the Salmonella/mammalian microsome assay. Environmental Mutagenesis 1:337

50. Maruoka S and Yamanaka S (1980) Production of mutagenic substances by chlorination of waters. Mutation Research 79:381

51. Kool HJ, Van Kreijl CF, Van Kranen HJ and de Greef E (1981) Toxicity assessment of organic compounds in drinking water in the Netherlands. Science of the Total Environment 18:135

52. De Marini DM, Pleura MJ and Brockman HE (1982) Use of four short-term tests to evaluate the mutagenicity of municipal water. Journal of Toxicology and Environmental Health 9:127

53. Grabow WOK, van Rossum PG, Grabow NA and Denkhaus R (1986) Relationship of the raw water quality to mutagens detectable by the Ames Salmonella/microsome assay in a drinking water supply. Water Research 15:1037

54. Meier JR, Lingg RD and Bull RJ (1983) Formation of mutagens following chlorination of humic acids. Mutation Research 118:25

55. Wilcox P and Horth H (1984) Microbial mutagenicity testing of water samples. In: Pascoe D and Edwards RW (eds) Freshwater Biological Monitoring. Pergamon, p 131

56. Fawell JK and Horth H (1990) Assessment and identification of genotoxic compounds in water. In: Waters MD et al. (eds) Genetic Toxicology of Complex Mixtures. Environmental Science Research Volume 39. Plenum, p 197

57. Van der Gaag MA, Noordsij A and Orange JP (1982) Presence of mutagens in Dutch surface waters and effects of water treatment processes for drinking water preparation. In: Mutagens in Our Environment. Alan R Liss, New York, p 277

58. Ringhand HP, Meier JR, Kopfler FC, Schenck KM, Kaylor WH and Mitchell DE (1987) Importance of sample pH on recovery of mutagenicity from drinking water by XAD resins. Environmental Science and Technology 21:382

59. Fawell JK, Fielding M, Horth H, James HA, Lacey RF, Ridgway JW, Wilcox P and Wilson I (1986) Health aspects of organics in drinking water. WRc Technical Report TR 231. Water Research Centre, Medmenham

60. Kringstad KP, Llungquist PO, De Sousa F and Stromberg LM (1983) On the formation of mutagens in the chlorination of humic acid. Environmental Science and Technology 17:553

61. Horth H (1989) Identification of mutagens in drinking water. Aqua 38:80

62 Holmbom B, Voss RH, Mortimer RD and Wong A (1978) Fractionation, isolation and characterisation of Ames mutagenic compounds in Kraft chlorination effluents. Environmental Science and Technology 18:333

63. Hemming J, Holmbom B, Reunanen M and Kronberg L (1986) Determination of the strong mutagen 3-chloro-4-(dichloromethyl)-5-hydroxy-2(5H)-furanone in chlorinated drinking and humic waters. Chemosphere 15:549

64. Kool HJ, Van Kreyl CF and Persad S (1989) Mutagenic activity in groundwater in relation to mobilisation of organic mutagens in soil. The Science of the Total Environment 84:185

65. Basu DK, Saxena J, Stoss FW, Santodanato J, Neal MW and Kopfler FC (1987) Comparison of drinking water mutagenicity with leaching of polycyclic aromatic hydrocarbons from water distribution pipes. Chemosphere 16:2595

66. Meier JR, Rudd CJ, Blazak WF, Riccio ES and Miller RG (1986) Comparison of the mutagenic activities of water samples disinfected with ozone, chlorine dioxide, monochloramine or chlorine. Environmental Mutagenesis 8:55

67. Huck PM, Anderson WB, Savage EA, Van Borstel RC, Daignault SA, Rector DW, Irvine GA and Williams DT (1989) Pilot scale evaluation of ozone and other drinking water disinfectants using mutagenicity testing. Ozone Science and Engineering 11:245

68. Hartemann P, Danglot C, Bourbigot MM, Pottenger L, Elias Z and Paquin JL (1987) Cellular toxicity assays and mutagenicity assays from on-site sampling of drinking water plants using multistage ozonation. Ozone Science and Engineering 9:179–194

69. Zoeteman BCJ, Hrubec J, de Greef E and Kool HJ (1982) Mutagenic activity associated with by-products of drinking water disinfection by chlorine, chlorine dioxide, ozone and UV-irradiation. Environmental Health Perspectives 46, p 197

70. Van Hoof F and Verheyden J (1981) Mutagenic activity in the river Meuse in Belgium 20:15

71. Van Rossum PG, Willemse JM, Hilner C and Alexander L (1982) Examination of a drinking water supply for mutagenicity. Water Science and Technology 14:163

72. Sato T, Ose Y, Nagase H and Hayase K (1987) Adsorption of mutagens by humic acid. The Science of the Total Environment 62:305

73. Noot DK, Anderson WB, Daignault SA, Williams DT and Huck PM (1989) Evaluating treatment processes with the Ames mutagenicity assay. Journal of the American Water Works Association 81:87

74. Monarca S, Pasquini R and Sforzolini GS (1985) Mutagenicity assessment of different drinking water supplies before and after treatment. Bulletin of Environmental Contamination and Toxicology 34:815

75. Van der Gaag MA, Kruithof JC and Puijker LM (1985) The influence of water treatment processes on the presence of organic surrogates and mutagenic compounds in water. The Science of the Total Environment 47:137

76. Wilcox P and Williamson S (1986) Mutagenic activity of concentrated drinking water samples. Environmental Health Perspectives 69:141

77. Wilcox P, Williamson S and Tye R (1990) Mutagenic activity of concentrated water extracts in cultured mammalian cells and Drosophila melanogaster. In: Jolley RL et al. (eds) Water Chlorination, Environmental Impact and Health Effects, Volume 6. Lewis, Michigan, p 239

78. Dunsire JP, Johnston AM and Paul HJ (1991) The disposition of [14C]-MX in the mouse. Report to Water Research Centre. In: The fate of the chlorination – derived mutagen MX in vivo. Foundation for Water Research. Report FR 0235, September 1991. FWR, Marlow, Bucks, UK

79. Varga C (1991) Genotoxicological evaluation of ozonated/chlorinated drinking water: cytogenetic effects of XAD-fractions on cultured human cells. Environmental Toxicology and Chemistry 10:1029

80. Tye RJ (1983) Comparison of Ames and mouse lymphoma LS 1787 test results on extracts derived from water samples. Environmental Mutagenesis 5:953

81. Lee PS, Rudd CJ and Meier JR (1986) Use of the mouse lymphoma mutagenesis assay to compare the activities of concentrates of treated drinking water. Environmental Mutagenesis 8 (Suppl 6):116

82. Wilcox P, Williamson S, Lodge DC and Bootman J (1988) Concentrated drinking water extracts which cause bacterial mutation and chromosome damage in CHO cells do not induce sex-linked recessive lethal mutations in *Drosophila*. Mutagenesis 3:381

83. Van der Gaag MA, Gauthier L, Noordsij A, Levi Y and Wrisberg MN (1990) Methods to measure genotoxins in wastewater: Evaluation with in vivo and in vitro tests. In: Waters MD et al. (eds) Genetic Toxicology of Complex Mixtures Environmental Science Research, Volume 30. Plenum, New York

84. Jaylet A, Gauthier L and Levi Y (1990) Detection of genotoxicity in chlorinated or ozonated drinking water using an amphibian micronucleus test. In: Waters MD et al. (eds) Genetic Toxicology of Complex Mixtures. Environmental Science Research Volume 39, Plenum, New York

85. Jaylet A, Gauthier L and Fernandez M (1987) Detection of mutagenicity in drinking water using a micronucleus test in newt larvae (*Pleurodeles watl.*). Mutagenesis 2:211

86. Fiskesjö G (1993) *Allium* Test 1: A 2–3 day plant test for toxicity assessment by measuring the mean root growth of onions (*Allium cepa L.*) Environmental Toxicology and Water Quality: An International Journal 8:461

87. Heartlein MW, De Marini DM, Katz AJ, Means JC, Pleura MJ and Brockman HW (1981) Mutagenicity of municipal water obtained from an agricultural area. Environmental Mutagenesis 3:519

88. Rank J and Nielsen MH (1994) Evaluation of the *Allium* anaphase – telophase test in relation to genotoxicity screening of industrial wastewater. Mutation Research 312:17

89. Ludyanskiy ML and Pasichny AP (1992) A system for water toxicity estimation. Water Research 26:689

Analysis of Organic Micropollutants in Drinking Water

R. A. Baumann and P. van Zoonen

RIVM, P.O. Box 1, 3720 BA Bilthoven, The Netherlands
E-mail: bert.baumann@rivm.nl

A vast number of organic micropollutants has been detected in the aqueous environment. A number of these compounds can eventually occur in our drinking water. The growing awareness of the risks associated with the occurrence of pollutants in drinking water led to the development of a diversity of sophisticated analytical methods. The basis of the major part of the currently used methodology is chromatography. Hyphenation of spectroscopic and chromatographic techniques is one of the most important trends in chromatography today. On the other end immunochemical techniques are emerging both as screening techniques and in combination with chromatography. This chapter gives an overview of the state-of-the-art in the analysis of the most important groups of pollutants in drinking water.

Keywords: micropollutants, analysis, drinking water, chromatography, pesticides.

Contents

The Handbook of Environmental Chemistry Vol. 5 Part C
Quality and Treatment of Drinking Water II (ed. by J. Hrubec)
© Springer-Verlag Berlin Heidelberg 1998

List of Abbreviations

AED	Atomic Emission Detector
APCI	Atmospheric Pressure Chemical Ionization
ASPEC	Automatic Sample Preparation with Extraction Columns
BTEX	Benzene, Toluene, Ethylbenzene and Xylene Isomers
DBP	Disinfection by-Product
EC	European Community
ECD	Electron Capture Detector
EEC	European Economic Community
ELISA	Enzyme Linked Immuno Sorbent Assay
EPA	Environmental Protection Agency
ETU	Ethylene Thio Urea
EU	European Union
FIIAA	Flow Injection Immuno Affinity Analysis
FMOC	9-FluorenylMethOxyCarbonyl
FPD	Flame Photometric Detector
GC	Gas Chromatography
GC-MS	Gas Chromatography-Mass Spectrometry
HPLC	High Performance Liquid Chromatography
LAS	Linear Aromatic Sulphonate
LC	Liquid Chromatography
LC-MS	Liquid Chromatography-Mass Spectrometry
L/L	Liquid-Liquid
LVI	Large Volume Introduction
MAC	Maximum Allowable Concentration
MCL	Maximum Contaminant Level
MS	Mass Spectrometry
MIMS	Membrane Inlet Mass Spectrometry
MRM	Multi Residue Method
NP	Nitrogen Phosphorous
PAH	Polycyclic Aromatic Hydrocarbon
PFBB	PentaFluoroBenzyl Bromide
PTV	Programmed Temperature Vaporization
SFC	Supercritical Fluid Chromatography
SFE	Supercritical Fluid Extraction
SPE	Solid Phase Extraction
SPME	Solid Phase Micro Extraction
SRM	Single Residue Method
UV	Ultra Violet
VOC	Volatile Organic Compound
WHO	World Health Organization

1
Introduction

Due to discharges from industrial production processes, a vast number of chemical compounds is continuously emitted to the environment. A high percentage of these chemicals is of an organic nature, with (physical) – chemical properties covering a wide range e. g. volatility, molecular weight, stability, hydrophobic nature etc. In agricultural production, among others, pesticides and fertilizers are applied in the open field, also leading to emissions to the environment.

Organic micropollutants, as defined in this chapter, are organic chemicals that are predominantly man-made and emerge in the environment due to the two pathways given above.

Since the early 1970s, the problems of environmental pollution have drawn the attention, first of small groups of environmentalists, while today it is a subject of wide public interest. The possible hazards related to environmental pollution with chemicals led to legislative actions concerning water and drinking water pollution. This resulted in the USA in the Safe Drinking Water Act in 1974, and in Europe in the Community drinking water directive (80/778/EEC) in 1980 and in the EC Priority Pollutants List in 1982. The EC Priority Pollutants List identifies hazardous substances, while the EC drinking water directive sets limits for 62 parameters in drinking water and water intended for drinking water production. These are either organoleptic, physico-chemical or microbiological parameters, or deal with organic or inorganic compounds. For pesticides the EC limit is 0.1 µg/l for a single pesticide and 0.5 µg/l for all pesticide residues. A study aimed at comparing the EC directive 80/778 with national legislations of EC Member States, other states and with WHO guidelines was prepared by Premazzi [1]. The issue of drinking water standards and regulations is reviewed in this Volume in detail by v. Dijk-Looijaard [2].

Through the legislation mentioned above and because of consumers actions initiated by concern for food and drinking water quality, an increasing demand for the analysis of contaminants in food and drinking water emerged. The increasing demand for analysis resulted in a considerable effort in analytical method development.

This chapter will focus on recent developments (approx. last 10 years) in analytical methodology and on a selection of organic micropollutants to analyse in drinking water and drinking water sources. Since EC directive 80/778 in particular led to expanding research activities in the pesticides field in Europe, a large part of this chapter will deal with analytical methods for pesticides. It is not aiming to be exhaustive in its text, but will address promising research. When reviews have been published recently on specific groups of compounds or on analytical techniques, these are referred to for more extensive references to the literature.

This introduction is followed by a section containing brief statements on interesting recent developments in analytical methodology. The chapter will further be built up in sections dealing with specific groups of organic micropollutants.

2
Developments in Analytical Methodology

The conventional approach for analyzing organic micropollutants in drinking water is (briefly) liquid-liquid (L/L) extraction of the compounds of interest with an organic solvent(mixture), concentration by evaporation, if necessary clean-up by column chromatography followed by concentration by evaporation, and gas chromatographic (GC) analysis. Generally speaking the analytical method consists of at least three steps that are performed independently and consecutively.

Improvements to this approach can be made in two different ways: either by replacing one (or more) step(s) in the analytical procedure by better one(s), or by integrating two or more steps to a one-step procedure.

An example of the former is the replacement of L/L extraction by solid phase extraction (SPE), extraction with Empore disks or solid phase micro extraction (SPME). The latter approach leads to the development of hyphenated techniques such as for instance an on-line combination of an autosampler capable of performing extraction and clean-up with a GC. The on-line approach leads to a reduction in the amount of manual operation and to a reduction of organic solvent consumption.

Another independent development was the establishment of reverse phase high performance liquid chromatography (HPLC) in the organic analytical laboratories, mainly due to the need to analyse polar pesticides that were emerging on the market. Due to its nature, reversed phase HPLC is fully compatible with aqueous samples.

The availability of relatively cheap, compact and robust mass spectrometers (MS) led to a displacement of GCs equipped with selective detectors (as the electron capture and the flame photometric detector) by GC-MS systems. With these systems it is possible to perform quantitative and/or confirming measurements. The development of interfaces for the coupling of LC to MS makes LC-MS a potentially interesting combination for the analysis of polar organic micropollutants in water. The atmospheric pressure chemical ionization (APCI) interface looks particularly promising in this respect.

Large volume injection (LVI) for GC by different techniques such as on-column injection by programmed temperature vaporization (PTV) makes it possible to introduce several hundreds of microliters onto a capillary GC column in stead of the typical 1 or 2 µl by conventional split-less or on-column injection techniques. This resulted in a two order reduction of the sample volume that had to be processed (500 ml to just a few ml) while maintaining the same sensitivity. This relatively small sample volume is much easier and faster to handle than the 0.5-l sample, leading to a more efficient sample preparation step.

Last but not least, analytical methods for organic micropollutants in water based on immunochemical principles were developed and became commercially available. At this moment immunoassays for about 35 pesticides and other micropollutants are on the market. Advantages of this technique are that hardly any sample handling is required, the analysis of samples can be performed simultaneously and relatively cheap. Also immunoaffinity chromatography where a specific antibody, immobilized on a stationary phase, selectively binds to the target analyte and simultaneously removes the analyte from the sample, is gaining attention.

Readers interested in an overview of analytical methods for organic micropollutants in water are referred to the book of Soniassy et al. [3]. This book gives extensive information about instrumentation and analytical procedures, mainly for (groups of) compounds occurring in the EC Priority Pollutants List. A number of the described methods is mandatory in the USA, as specified by the US Environmental Protection Agency (EPA). In *Analytical Chemistry*, bi-annual literature reviews on water analysis are published, the two most recent ones are listed here [4, 5]. Lindner [6] described analytical methods for the determination of a group of compounds (e.g. alcohols, amines, carboxylic acids and aromatic sulfonates) in waste water and river water at the (sub) µg/l level.

Ferguson et al. [7] gave an overview of the use of immunochemical techniques for the trace analysis of pesticides in environmental and agricultural samples. Sherry [8] reviewed immunochemical methods for environmental analysis.

3
Organic Micropollutants Other than Pesticides

3.1
Volatiles

The group of volatile organic compounds (VOCs) comprises low molecular weight aliphatic and aromatic (halogenated) hydrocarbons, alcohols, ketones, nitriles and carboxylic acids. Some of the most frequently occurring volatile organic pollutants in drinking water are halogenated hydrocarbons [9], especially the trihalomethanes ($CHCl_3$, $CHCl_2Br$, $CHClBr_2$ and $CHBr_3$), which are the most important group of drinking water disinfection by-products (DBPs). Under the conditions of drinking water treatment practice, DBPs in the concentration range of hundreds of µg/l can be formed.

Another important group of VOCs of high health risk are the aromatic hydrocarbons, like benzene and alkylated benzenes. The phenols, a group of compounds that contains several VOCs, are discussed separately in Sect. 3.3.

Because of the high volatility of VOCs, proper sampling and preservation procedures have to be maintained. An advantage of their volatility is that they can easily be removed from water samples with various techniques like purging with a gas, while potentially interfering compounds stay in the water, adding extra selectivity to the overall analytical procedure. The state-of-the-art in the analysis of VOCs in water and sediments has been reviewed recently [10].

A well established method for the sampling of volatile organic compounds from water is the static headspace method. With this method an equilibrium is reached between components in the water sample and the gas phase (headspace), some time after the sample is brought in a closed vessel. An aliquot of the headspace is injected onto the gas chromatographic system. Poy and Cobelli

[11] described a system where an automatic static headspace sampler was coupled to a GC equipped with a programmed temperature vaporizer injector. With this system a large amount of headspace can be rapidly transferred to the GC without efficiency loss.

With the dynamic headspace technique, the gaseous phase above the sample is continuously purged with a gas, which carries the analytes away. After this, the analytes are trapped on a solid sorbent like charcoal, or in a liquid solution. When the gaseous phase is led through in stead of over the sample, this is called the purge and trap method. Compared to the static headspace method, with both dynamic methods compounds with a lower volatility can also be determined. Moreover lower limits of determination can be achieved, because a large fraction of the analytes can be removed from a sample. In a paper by Abeel et al. [12], developments in the purge and trap technique were reviewed. A purge and trap system for the determination of a large group of VOCs in drinking water was described by Ho [13].

A new, interesting approach for the removal of VOCs from water samples is membrane extraction. With this technique a water sample is brought into contact with a membrane which selectively permeates compounds to a gas phase. The analytes are directly transferred to the chromatographic system for analysis, or first trapped on an absorbent and transferred later. It is even possible to couple a membrane directly to a mass spectrometer, without the GC column. This technique, termed membrane inlet mass spectrometry (MIMS) was applied for the determination of trihalomethanes and other VOCs in chlorinated drinking water at the 0.1 µg/l level [14].

Solid phase micro extraction (SPME) is a novel extraction technique based on the adsorption of analytes from a sample on a polymer coated fiber. SPME can be used for gaseous as well as aqueous samples combined with GC or GC-MS systems, because the fiber can be inserted in a GC injection port [15]. Advantages of the method are its simplicity and speed and the fact that it has no solvent consumption. However, in order to obtain reliable results much effort has to be put in calibration. Potter and Pawliszyn [16] developed a method for the determination of BTEX (benzene, toluene, ethylbenzene and the xylene isomers) in water by SPME in combination with GC-MS at the µg/l level. In Fig. 1 an SPME device is shown. Nilsson et al. [17] reported on an inter-laboratory study for the quantitative analysis of 13 VOCs in drinking water with SPME as the extraction method. SPME was compared with the purge and trap and with the static headspace technique by analysis of a reference sample. It was concluded that SPME is an accurate and precise extraction technique suited for quantitative routine analysis of VOCs in drinking water at trace level.

3.2
Polycyclic Aromatic Hydrocarbons

A substantial part of polycyclic aromatic hydrocarbons (PAHs) present in the environment originates from incomplete combustion processes. Due to this formation route, they are widespread in all compartments of the environment. Some PAHs, like benzo(a)pyrene and benzo(b)fluoranthene, are carcinogenic.

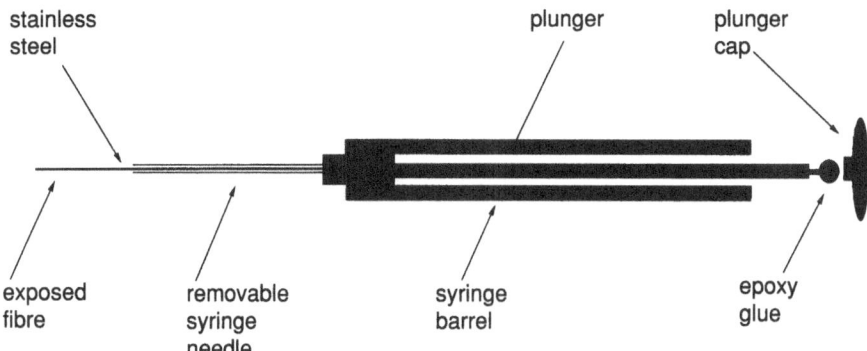

Fig. 1. A solid phase micro-extraction device. (Reprinted from J Chromatogr 625, Potter DW and Pawliszyn J (1992), p. 248, with kind permission of Elsevier Science – NL, Sara Burgerhartstraat 25, 1055 KV Amsterdam, The Netherlands)

The 16 so-called EPA PAHs are considered priority pollutants, they are extensively monitored in drinking water. The standards and guidelines for PAHs in drinking water have been reviewed in the Handbook elsewhere [2].

Until recently PAHs were preferably analyzed by HPLC with fluorescence detection, probably due to the fact that in gas chromatography there was no selective detector available for the analysis of these compounds. The introduction of cheap and reliable bench top mass spectrometers revived the interest for GC-MS detection of PAHs.

Bulman [18] reviewed on the determination of PAHs in environmental water samples.

Matthiessen [19] developed a selective extraction method of PAHs from water samples by sorption onto a solid phase containing nitro groups, due to the formation of charge-transfer complexes between the PAHs and the electron acceptor on the solid phase.

Janssen et al. [20] analyzed PAHs in the (sub) µg/g concentration range in river sediment with a GC-MS equipped with a large volume injection system. An organic solvent extract of the sediment was purified using solid phase extraction. With the OPTIC injector, 50 µl purified extract were injected onto the capillary GC column.

Gremm and Frimmel [21] reported on an LC-MS method employing the particle beam interface for the quantification and identification of PAH metabolites in water obtained from a biological degradation experiment. Evaluation of spectral data indicated the presence of six main PAH metabolites. The application of microbore LC separation columns proved advantageous. LC-MS results where comparable with results obtained by GC-MS.

3.3
Phenols

Phenols emerge in the environment due to such different pathways as industrial emissions and the biodegradation of natural compounds. Chlorinated phenols are formed during the chlorination of drinking water. Some of them are considered as toxic and many of them have a very low (µg/l) odour and taste threshold concentration [22]. Limits for these compounds in drinking water are based on either organoleptic or toxicological (e.g. pentachlorophenol) considerations.

Marko-Varga [23] reviewed on analytical methods for phenols in water. It is not quite clear whether there is a preferred method of analysis for phenols in current practice, e.g. GC with derivatization or LC. For the analysis of phenols in drinking water all methods will probably suffice, for surface water and waste water GC is preferred because of its higher selectivity.

Schmidt et al. [24] reported on the use of membrane extraction disks followed by GC or LC for the determination of phenols in aqueous samples. In Table 1 recovery values for a group of phenols adsorbed from different water samples by using a thick membrane loaded with acetyl-derivatized resin are shown.

A comparative study on the performance of a number of solid sorbents for the extraction of priority phenolic compounds from environmental waters was described by Puig and Barcel [25]. They found that breakthrough volumes for the selected phenols and the selectivity of the extraction were dependent on the type of sorbent.

SPME followed by GC analysis was applied for the determination of phenol in water at the sub µg/l level [26].

3.4
Phthalates

Phthalates are diesters of phthalic acid, 1,2-benzenedicarboxylic acid. They are used as softeners in PVC products. Phthalate production is estimated at several tenths of million tons up till now. Phthalates are reported to have reproductive and endocrine-disrupting effects [27]. They are ubiquitous, implying that special care has to be taken to prevent contamination during sampling and analysis. Because of this ubiquity, analytical methods that require a minimum of sample handling and a minimum of solvent use are to be preferred. There are not many references to be found in the open literature on the analysis of phthalates in water.

Ritsema et al. [28] described a method based on solid phase extraction of the phthalates from surface water followed by GC-ECD and GC-MSD analysis. The limit of determination in river water was between 0.01 and 0.1 µg/l; the method was applied for a 12-day survey of phthalates in water and sediment from the river Rhine.

3.5
Surfactants

Surfactants are a diverse group of compounds exhibiting hydrophilic and/or hydrophobic properties, depending on their (physico)-chemical environment. They are widely in use, both for industrial and domestic applications, and thus large amounts are emitted to the environment, making them ubiquitous. The

Table 1. Recovery of phenols using thick membranes loaded with acetyl-derivatized resin in a small tube

Elution with 0.75 ml methanol. Results are the average of three individual results.

Analyte	Distilled water Recovery (%)	R.S.D. (%)	Tap water Recovery (%)	R.S.D. (%)	River water Recovery (%)	R.S.D. (%)
Phenol	99	2	96	2	97	3
3-Methylphenol	95	1	93	2	92	4
2-Nitrophenol	100	4	102	3	98	1
2,4-Dichlorophenol	98	0	97	3	97	2
4-Chloro-3-methylphenol	98	0	85	2	93	4
2,4,6-Trichlorophenol	102	0.5	98	3	101	0
2-Methyl-4,6-dinitrophenol	94	2	75	4	95	2
Pentachlorophenol	100	2	92	1	100	4
2-Chlorophenol	99	2	106	4	103	2
2-Methylphenol	99	4	100	3	100	1
4-Methylphenol	97	5	98	2	98	4
2,4-Dimethylphenol	97	5	94	2	103	5
2,6-Dichlorophenol	100	5	108	8	104	6
2,4,5-Trichlorophenol	96	5	95	5	107	3
2,3,4,6-Tetrachlorophenol	99	3	95	9	101	0.5
2-sec-Butyl-4,6-dinitrophenol	101	4	97	4	101	2

estimated annual world consumption of surfactants at present is estimated at 15 million metric tons.

EC directive 80/778 fixes a MAC value for surfactants of 200 µg/l. It is noted that the concentration of detergents in drinking water should not be allowed to reach levels giving rise to either foaming or taste and odor problems.

Within the group of surfactants, three subgroups can be distinguished: cationic (quaternary ammonium salts), anionic (sulfonates) and non-ionic (most important groups alkylphenol ethoxylates and alcohol ethoxylates). Alkylphenol polyethoxylates and the degradation product nonylphenol are demonstrated to have a weakly estrogenic effect [29].

Most non-ionic surfactants are mixtures of homologs and/or isomers; the actual composition is dependent on conditions maintained during the synthesis process. This composition, and the fact that several groups of surfactants may be present in one sample, makes it necessary to apply powerful separation techniques for their analysis.

Kiewiet and de Voogt [30] reviewed chromatographic methods of analysis for non-ionic surfactants (alkylphenol ethoxylates and alcohol ethoxylates) in the aquatic environment. The authors stated that at present solid phase extraction, including chromatographic clean-up, followed by liquid chromatography (normal phase or reversed phase) with UV or fluorescence detection is the method of choice for the determination of the mentioned non-ionic surfactants at the µg/l level in aqueous samples. High temperature GC-ECD or -MS can also be applied as analytical method, but a disadvantage of this technique is that sample degradation for the higher molecular mass surfactants occurs.

At present, liquid chromatography coupled on-line to mass spectrometry (LC-MS) is the most sensitive and selective method of analysis for non-ionic surfactants. It enables the determination of single homologs and isomers in water at the µg/l level. Moreover, the information obtained with LC-MS can be used to identify possible sources of contamination with surfactants. Evans et al. [31] developed a method based on thermospray LC-MS for analysing alcohol ethoxylates in water. In Fig. 2 thermospray LC-MS spectra for two compounds are shown.

An interesting approach in the analysis of non-ionic surfactants is the on-line combination of supercritical fluid extraction and supercritical fluid chromatography (SFE-SFC) [32]. SFE-SFC was applied to their extraction and analysis in aqueous samples.

The most common anionic surfactants are the linear alkylbenzene sulfonates (LAS). Reemtsma [33] reviewed methods of analysis for polar aromatic sulfonates in aqueous samples. The routine method for analysing LAS in aqueous samples comprises extraction, clean-up and reversed phase HPLC. Sulfonates that are very polar, due to substitution with specific groups or due to polysulfonation, can successfully be analyzed by ion-pair HPLC. In both cases, the HPLC system may be equipped with a UV or a fluorescence detector. Scullion et al. [34] described the determination of LAS in surface water at the µg/l level by liquid chromatography with fluorescence detection and LC-APCI-MS respectively. This method allows the determination of individual LAS homologs in a water sample in the presence of non-ionic surfactants.

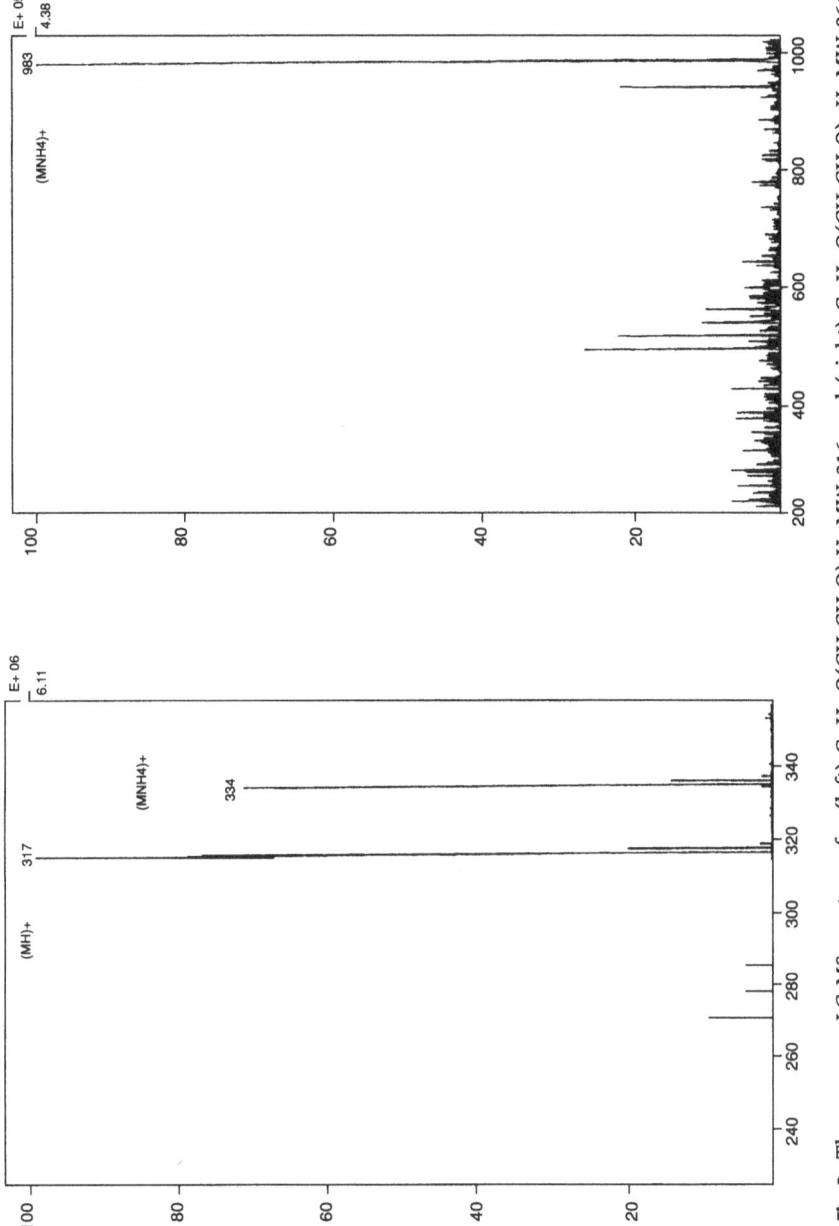

Fig. 2. Thermospray LC-MS spectrum for (left) $C_{15}H_{31}O(CH_2CH_2O)_2H$, MW 316, and (right) $C_{11}H_{23}O(CH_2CH_2O)_{18}H$, MW 964. (Reprinted with permission from KA Evans et al., Anal Chem 66:701. Copyright (1994) American Chemical Society)

4
Pesticides

4.1
Introduction

Pesticides are intensively used in agri- and horticultural production, along rail-way tracks and roads, in households and in cooling systems. During these applications, they are also emitted to the environment. There are two main routes through which pesticide residues can occur in sources intended for drinking water production. The first is via surface water, where they emerge due to spray drift after application in the field and due to atmospheric deposition. Another source of pesticides in surface water is waste water from production processes. The second is via ground water, where they emerge due to leaching from soil. Treatment steps currently in use in the production process of drinking water, like charcoal filtration, are not able to remove all pesticide residues. Therefore, monitoring of drinking water and drinking water sources for pesticides, is necessary. Some drinking water standards as EC drinking water directive 80/778 sets limits for a single pesticide, and for all pesticide residues at the sub µg/l concentration level [2]. These very low limits are not based on toxicologically save values, but are based on the assumption that pesticides should not be present in drinking water at all. The sub µg/l concentration level is the lowest limit of determination possible to achieve with state-of-the art analytical methods for a number of pesticides in the time EC 80/778 was made up. In the United States the Safe Drinking Water Act [35], passing U.S. Congress in 1974, requires EPA to regulate those contaminants that may have an adverse effect on human health. For those compounds maximum contaminant levels (MCLs) are established. Depending on toxicological data, MCLs for specific pesticides in drinking water thus range from the ng/l to the mg/l level. The World Health Organization (WHO), in its 1984 Guidelines for drinking water quality [36] states that values for constituents must be based on known health related data, such that the water is suitable for human consumption over a lifetime.

The selection of publications from the literature referred to in this section, takes into account that the described method should be able to analyze the pesticides at the 0.1 µg/l concentration level.

Recently, Stan [37] published two volumes on the analysis of pesticides in ground and surface water. Volume I mainly reviews existing methodology, while volume II deals with latest developments in the field. Hatrik and Tekel [38] reviewed extraction methodology and chromatography for the determination of residual pesticides in water. They emphasized on multi residue (e.g. more than one compound) analytical methods with the required limit of quantification of 0.1 µg/l.

Readers who are interested in analytical methods for the determination of residues of pesticides in a wide variety of sample matrices are referred to the bi-annual literature reviews of Sherma in *Analytical Chemistry*, the two most recent ones are listed here [39, 40].

4.2
Pesticides Amenable to Gas Chromatography

Due to its separation power capillary gas chromatography (GC), in combination with a selective and sensitive GC detector, still is the analytical method of choice for a considerable number of pesticides. GC is suited for the direct determination of relatively apolar, thermostable compounds in single (SRM) or multi residue methods (MRM). Another group of pesticides can be analyzed with GC after derivatization with a suitable reagent. Examples of groups of compounds that can successfully be analyzed directly with GC are the organochlorine pesticides, organophosphorous esters and triazines.

Sandra et al. [41] compared the performance of solid phase extraction (SPE) on cartridges and on extraction disks for the extraction of triazines from water samples. Analysis was by GC-MS equipped with a PTV injector. They stated that extraction disks are promising because of their fast sampling rate and low solvent consumption.

Also for pesticides, solid phase micro extraction (SPME) is a new promising extraction technique. The extraction efficiency of the polymer coating of the SPME fiber strongly depends on the matching of the polarity of the polymer and the polarity of the specific pesticide. This means that for the extraction of relatively (non-)polar compounds, a relatively (non-)polar polymer coating should be chosen.

Eisert and Levsen [42] combined SPME extraction with analysis on a GC equipped with an Atomic Emission Detector (AED) for organophosphorous pesticides in aqueous samples. Using a very small sample volume (3–10 ml) they were able to analyze the compounds at the 0.1 µg/l level. A number of pesticides, including triazines, where analyzed with GC-MS after SPME (extraction time 45 min) on a polydimethylsiloxane fiber [43].

A review on trace level analysis of micropollutants in aqueous samples using GC with on-line sample enrichment and large volume injection was published recently by Mol et al. [44].

Van der Hoff et al. [45] developed a fully automated method for the determination of pyrethroids in surface and drinking water at the ng/l level with solid phase extraction coupled to large volume GC-ECD. The solid phase extraction cartridges, filled with silica, were applied for a combined extraction/clean-up step. The same author described an on-line combination of automated micro L/L extraction and capillary GC equipped with a flame photometric detector (FPD) for the determination of a group organophosphorous pesticides in water [46]. A volume of 1.5 ml sample was extracted with 1.5 ml methyl *tert*-butyl ether. A volume of 500 µl extraction solvent was injected onto the capillary column via an on-column injection technique. Recoveries for the pesticides were >70 % in most cases, the limit of quantification was 0.1 µg/l.

Goosens et al. [47] investigated the on-line coupling of L/L extraction with capillary GC for the determination of hexachlorocyclohexanes at the 0.1 µg/l level in groundwater, employing a sandwich phase separator. Hankemeier et al. [48] developed an automated at-line SPE-GC procedure for pesticides and other micropollutants in water using a semi-robotic autosampler. In Fig. 3, a schematic of the coupled autosampler-GC system is given.

Noij and van der Kooi [49] reported on a fully automated analysis for nitrogen and phosphorous (NP) containing pesticides in water by on-line SPE-GC. With this set-up the instrument can work unattended for the analysis of over 50 samples on 33 relatively polar NP pesticides at concentration levels between 0.01 and 0.1 μg/l.

An on-line approach for the coupling of sample preparation and GC-MS for the analysis of some triazines in tap water was described by Janssen et al. [20]. An ASPEC, an apparatus able to perform automated solid phase extraction with SPE cartridges, was coupled to an OPTIC injector equipped on the GC-MS system. Starting with a 1 ml water sample, 100 μl ethyl acetate extract obtained from the ASPEC were injected on-line onto the capillary GC column by way of the OPTIC. The triazines could be determined in tap water at the μg/l concentration level.

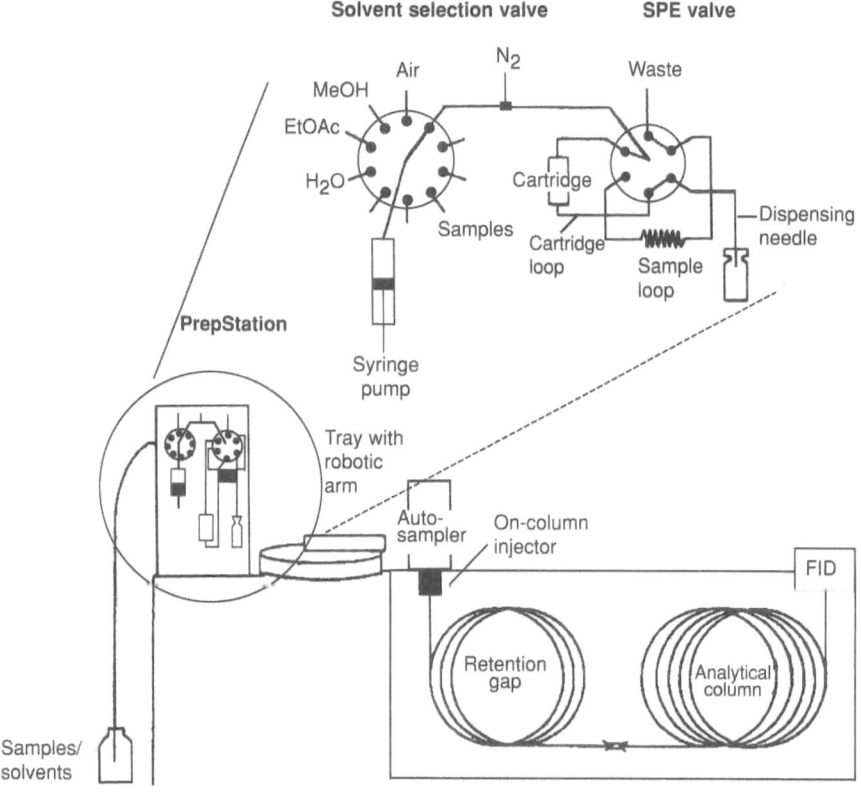

Fig. 3. Set-up of the autosampler-GC-FID system. Abbreviations: EtOAc, ethyl acetate; H₂O, water; MeOH, methanol; N₂, nitrogen. (Reprinted from J Chromatogr A 750, Hankemeier Th, Steketee PC, Vreuls JJ, Brinkman UATh (1996), p. 162, with kind permission of Elsevier Science – NL, Sara Burgerhartstraat 25, 1055 KV Amsterdam, The Netherlands)

Meiring and de Jong [50] described a method for the determination of ethyl-enethiourea (ETU) in water by single-step extractive derivatization and GC-MS with negative ion chemical ionization. The method was applied for confirmation purposes for the presence of ETU in samples as analyzed by HPLC-UV. In general, a good correlation was found between the results from both methods. Vink and van der Poll [51] reported on a multi residue method for eight polar acidic herbicides in surface water. The compounds were derivatized with pentafluorobenzyl bromide (PFBB), and the derivatives were analyzed with GC equipped with an electron capture detector (ECD) and GC-MS at the 0.02–0.05 µg/l level.

4.3
Pesticides Amenable to Liquid Chromatography

Liquid chromatography (LC) is used for the determination of pesticides, and metabolites, which cannot be analyzed directly by GC due to low volatility, high polarity and/or thermal instability. Applying GC analysis to these sort of compounds often requires a complicated derivatization procedure prior to GC separation. Examples of groups of compounds that can be analyzed successfully with LC are the chlorophenoxy acids, N-methylcarbamates, phenylureas and sulfonylureas.

Pichon and Hennion [52] developed a method for the determination of pesticides in environmental waters by automated on-line trace enrichment and liquid chromatography with UV detection. The choice of sorbent and amount of sample volume were discussed. Detection limits of 0.1 µg/l were obtained using 150 ml of river water without any additional clean-up.

Hogendoorn [53] investigated strategies in method development for the determination of polar pesticides and metabolites with coupled column liquid chromatography (LC-LC). An important aspect of the application of coupled column LC is the possibility to integrate sample preparation and clean-up on-line in the chromatographic procedure. For certain compounds like bentazone and the phenyl urea herbicides, which have favorable chromatographic and UV absorbing properties, it is even possible to perform direct analysis at the 0.1 µg/l level in water by large volume injection [54].

Kiso et al. [55] described the analysis of a group of 15 pesticides at the (sub) µg/l level in environmental water samples by direct sample injection by isocratic and gradient elution with reversed phase LC and UV detection.

Sancho et al. [56] reported on the determination of glufosinate in environmental water samples using precolumn derivatization with 9-fluorenyl-methoxycarbonyl (FMOC), large volume injection and coupled column LC. The used HPLC set-up is illustrated schematically in Fig. 4. Barcèlo [57] edited a book on applications of LC-MS in environmental chemistry. An interesting contribution to this book is by Geerdink [58] on flow injection analysis coupled to thermospray tandem mass spectrometry.

Honing et al. [59] developed a method for the determination of carbamate pesticides in water and sediment samples by LC-MS, with both the ionspray and

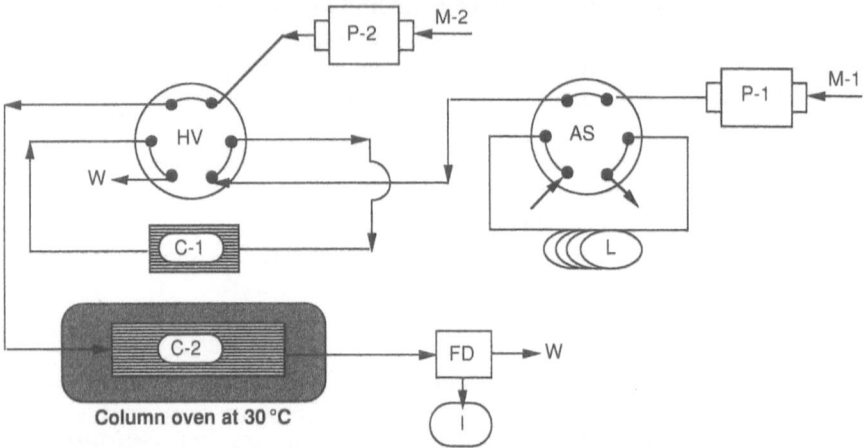

Fig. 4. HPLC set-up for column-switching. *AS*=sample injector with a 2 ml loop (L); *HV*=six-port high-pressure valve; *P-1*=gradient LC pump; *P-2*=isocratic LC pump; *C-1*=first separation column; *C-2*=second separation column; *M-1* and *M-2*=mobile phases on C-1 and C-2, respectively; *FD*=fluorescence detector; *I*=integrator system; *W*=waste. (Reprinted from J Chromatogr A 678, Sancho JV, López FJ, Hernández F, Hogendoorn EA, van Zoonen P (1994), p. 61, with kind permission of Elsevier Science – NL, Sara Burgerhartstraat 25, 1055 KV Amsterdam, The Netherlands)

the thermospray interface. The ionspray interface provided better sensitivity and selectivity than the thermospray interface. Application of the ionspray interface permitted the determination of the carbamates at the 0.1 µg/l level providing a large volume injection of 0.5 ml was performed.

Hogendoorn et. al [60] described a unified approach for the analysis of polar pesticides in water utilizing coupled column LC for screening followed by GC, after derivatization when necessary, for confirmation. The results demonstrate that the proposed LC and MS techniques are capable of analyzing a wide range of polar pesticides down to a level of 0.1 µg/l.

4.4
Immunochemical Techniques for Pesticide Analysis

All immunochemical techniques are based on the specific binding between an antibody and the target analyte. Depending on the specific format this binding reaction can be performed in a small volume of aqueous solution, in a flow injection system or in a liquid chromatographic system. In the last two cases the antibody is immobilized on a packing material like agarose or acrylate gels, or silica.

In the first case the applied technique is the immunoassay. During the last 10 years, the number of commercially available enzyme linked immunosorbent assays (ELISAs) for pesticide analysis has steadily increased. They provide simple, selective and rapid analytical methods; at present assays for about 30 different pesticides are on the market. Due to the selectivity of the immuno-

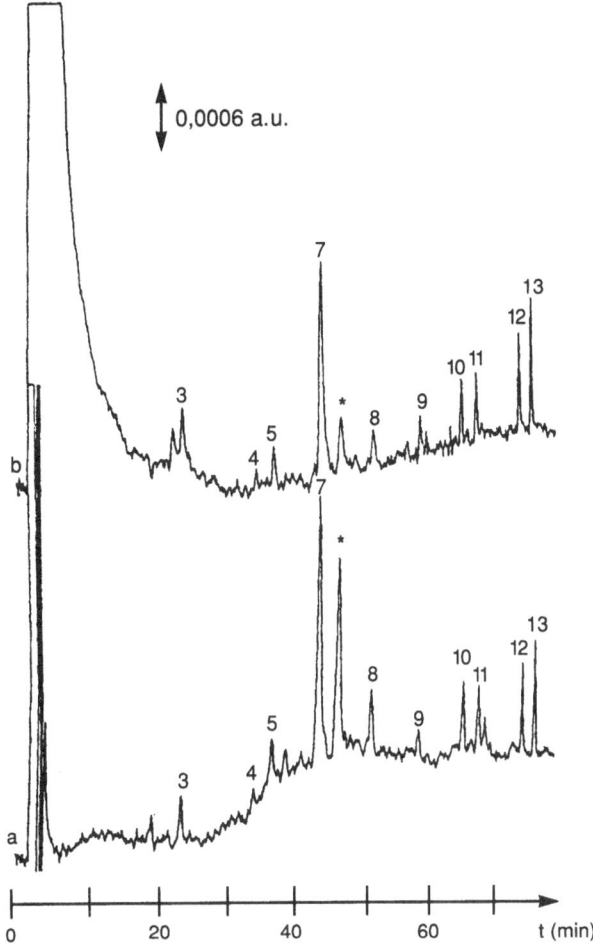

Fig. 5. On-line analysis of 50 ml drinking water (a) and river Seine water (b) spiked at 0.5 µg/l with 13 phenylureas. (*1*) fenuron, (*2*) methoxuron, (*3*) monuron, (*4*) methabenzthiazuron, (*5*) chlortoluron, (*6*) fluometuron, (*7*) isoproturon, (*8*) difenoxuron, (*9*) buturon, (*10*) linuron, (*11*) chlorbromuron, (*12*) diflubenzuron and (*13*) neburon. * = impurity originating from the synthesis of the immunosorbent (Reprinted from Anal Chim Acta 311, Pichon V, Chen L, Hennion MC (1995), p. 432 and 434, with kind permission of Elsevier Science – NL, Sara Burgerhartstraat 25, 1055 KV Amsterdam, The Netherlands)

chemical binding reaction, these methods in general require no or simple sample handling.

Hock et al. [61] reviewed immunochemical techniques and antibody production for pesticide analysis. Mouvet et al. [62] evaluated commercially available ELISA microtiter plates for triazines in water samples. Quality control of river water intended for the production of drinking water by the application of immunoassays for the detection of the chlorophenoxy acid 2,4-D was reported by Meulenberg and Stoks [63]. Comparison with a GC-MS reference method revealed no false negative results and a rate of false positive results of about

10%. The authors concluded that the assay could be applied in the early warning.

Krämer et al. [64] reported on a newly developed immunochemical detection system for the determination of pesticides in water. The technique is called flow injection immuno affinity analysis (FIIAA) and is based on the specific binding of antibodies, immobilized on an protein A affinity column, with the compound of interest. The system was tested for the analysis of atrazine and will also be tested in the future for diuron. The comparison of FIIAA with ELISA results for atrazine was in good agreement in most cases.

Pichon et al. [65] developed a method based on on-line pre-concentration and LC analysis for 12 phenylurea hebicides in environmental water using a silica based immunosorbent. The immunosorbent was packed in a conventional LC pre-column. Using 50 ml river water detection limits of between 0.05 and 0.5 µg/l were found; Isoproturon was monitored in River Seine water. Figure 5 shows two chromatograms of the on-line analysis of drinking water and river water respectively, spiked at 0.5 µg/l with 13 phenylurea herbicides. The same research group reported on the immunoaffinity technique for the analysis of triazines in river water [66, 67].

5
Conclusion

In the last decade, considerable progress has been made in the development of reliable and cost-effective analytical techniques for the assessment of water quality. Automation can be accomplished by the use of coupled chromatographic techniques, while detection is improved by the hyphenation of chromatographic and spectroscopic techniques. Developments in other areas such as immunochemistry provides new approaches towards screening and sample preparation.

References

1. Premazzi G, Chiaudani G, Ziglio G (1989) Scientific assessment of EC standards for drinking water quality, Report EUR 12427, JRC-Ispra and DGXI, Office for Official Publications of the European Communities, Luxembourg
2. van Dijk-Looijaard AM (1995) In: Hrubec J (ed). Quality and treatment of drinking water, the Handbook of Environmental Chemistry, vol 5B, Springer, Berlin Heidelberg New York, p 3
3. Soniassy R, Sandra P, Schlett C (1994) HP Part No. 5962–6216E, Hewlett-Packard Company, Waldbronn, Germany
4. MacCarthy P, Klusman RW, Cowling SW, Rice JA (1993) Anal Chem 65:244R
5. MacCarthy P, Klusman RW, Cowling SW, Rice JA (1995) Anal Chem 67:525R
6. Lindner K (1996) Erfassung und Identifizierung von trinkwassergängigen Einzelsubstanzen in Abwässern und im Rhein. ARW/VCI-Forschungsvorhaben, Köln/Frankfurt
7. Ferguson BS, Kelsey DE, Fan TS, Bushway RJ (1993) Sci Total Environ 132:415
8. Sherry JP (1992) Crit Rev Anal Chem 23:217
9. Versteegh JFM, van Gaalen FW, Beuting DM (1996) The quality of drinking water in the Netherlands in 1994, RIVM report no. 731011010, National Institute of Public Health and the Environment, Bilthoven, The Netherlands (in Dutch)
10. Kurán P, Soják L (1996) J Chromatogr A 733:119

11. Poy F, Cobelli J (1985) J Chromatogr Sci 23:114
12. Abeel SA, Vickers AK, Decker D (1994) J Chromatogr Sci 32:328
13. Ho JS (1989) J Chromatogr Sci 27:91
14. Bauer SJ, Solyom D (1994) Anal Chem 66:4422
15. Lough D, Mostlagh S, Pawliszyn J (1992) Anal Chem 64:844A
16. Potter DW, Pawliszyn J (1992) J Chromatogr 625:247
17. Nilsson T, Ferrari R, Facchetti S (1996) In: Sandra P (ed) Proc 18th Int Symp on Cap Chrom, Hütig Verlag, Heidelberg, Germany, p 618
18. Bulman TL (1992) Environ Sci Pollut Control Ser 4:149
19. Matthiessen A (1992) Vom Wasser 79:159
20. Janssen HG, Perkins P, Hutchinson G, Fraudeau C (1996) Int Env Technol 6 (3):14
21. Gremm TJ, Frimmel FH (1994) Chromatographia 38:781
22. Mallevialle J, Bruchet A (1995) In: Hrubec J (ed) Quality and treatment of drinking water, the Handbook of Environmental chemistry, vol 5B, Springer, Berlin Heidelberg New York, p 139
23. Marko-Varga GA (1993) Tech Instrum Anal Chem 13:225
24. Schmidt L, Sun JJ, Fritz JS, Hagen DF, Markell CG, Wisted J (1993) J Chromatogr 641:57
25. Puig D, Barcel D (1996) J Chromatogr 733:371
26. Buchholz KD, Pawliszyn J (1993) Environ Sci Technol 27:2844
27. Jobling S, Reynolds T, White R, Parker MG, Sumpter JP (1995) Environm Health Perspect 103:582
28. Ritsema R, Cofino W, Frintrop PCM, Brinkman UATh (1989) Chemosphere 18:2161
29. Jobling S, Sumpter JP (1993) Aquat Toxicol 27:361
30. Kiewiet AT, de Voogt P (1996) J Chromatogr A 733:185
31. Evans KA et al. (1994) Anal Chem 66:699
32. Kane M, Dean JR, Hitchen SM, Dowle CJ, Tranter RL (1993) Anal Proc 30:399
33. Reemstsma T (1996) J Chromatogr A 733:473
34. Scullion SD, Clench MR, Cooke M, Ashcroft AE (1996) J Chromatogr A 733:207
35. Environmental Protection Agency (1974) U.S. Public Law 93–523, 93rd Congress, p 433 Safe Drinking Water Act, U.S. Washington DC
36. World Health Organization (1993) Guidelines for drinking water quality, Geneva, Switzerland
37. Stan HJ (1995) Analysis of pesticides in ground and surface water, vols I and II, 1st edn. Springer, Berlin Heidelberg New York
38. Hatrik S, Tekel J (1996) J chromatogr A 733:217
39. Sherma J (1993) Anal Chem 65:40R
40. Sherma J (1995) Anal Chem 67:1R
41. Sandra P, Haghebaert K, David F (1996) Int Environ Technol 6 (2):13
42. Eisert R, Levsen K (1994) In: Sandra P (ed) Proc 18th Int Symp on Cap Chrom, Hütig Verlag, Heidelberg, Germany, p 544
43. Sandra P, Haghebaert K, David F (1996) Int Environ Technol 6 (5):6
44. Mol HGJ, Janssen HGM, Cramers CA, Vreuls JJ, Brinkman UATh (1995) J Chromatogr A 703:265
45. Van der Hoff GR, Pelusio F, Brinkman UATh, Baumann RA, van Zoonen P (1996) J Chromatogr A 719:59
46. Van der Hoff GR, Baumann RA, Brinkman UATh, van Zoonen P (1993) J Chromatogr 644:367
47. Goosens EC, Bunscoten RG, Engelen G, de Jong D, van den Berg JHM (1990) J High Resolut Chromatogr 13:438
48. Hankemeier Th, Steketee PC, Vreuls JJ, Brinkman UATh (1996) J Chromatogr A 750:161
49. Noij ThHM, van der Kooi MME (1995) J High Res Chromatogr 18:535
50. Meiring HD, de Jong APJM (1994) J Chromatogr A 683:157
51. Vink M, van der Poll JM (1996) J Chromatogr A 733:361
52. Pichon V, Hennion MC (1994) J Chromatogr A 665:269
53. Hogendoorn EA (1993) PhD thesis, Free University Amsterdam, The Netherlands

52   R. A. Baumann and P. van Zoonen

54. Hogendoorn EA, Brinkman UATh, van Zoonen P (1993) J Chromatogr 644:307
55. Kiso Y, Li H, Shigetoh K, Kitao T, Jinno K (1996) J Chromatogr A 733:259
56. Sancho JV, López FJ, Hernández F, Hogendoorn EA, van Zoonen P (1994) J Chromatogr A 678:59
57. Barcélo D (1996) Applications of LC-MS in environmental Chemistry, 1st edn. Elsevier, Amsterdam
58. Geerdink RB (1996) Optimization of instrumental parameters for flow-injection analysis, thermospray tandem mass spectrometry. In: Barcélo D (ed) Applications of LC-MS in environmental chemistry, 1st edn. Elsevier, Amsterdam, Ch 3
59. Honing M, Riu J, Barcélo D, van Baar BLM, Brinkman UATh (1996) J Chromatogr A 733:283
60. Hogendoorn EA, Hoogerbrugge R, Baumann RA, Meiring HD, de Jong APJM, van Zoonen P (1996) J Chromatogr A 754:49
61. Hock B, Dankwardt A, Kramer K, Marx A (1995) Anal Chim Acta 311:393
62. Mouvet C, Broussard S, Jeannot R, Maciag C, Abuknesha R, Ismail G (1995) Anal Chim Acta 311:331
63. Meulenberg EP, Stoks PG (1995) Anal Chim Acta 311:407
64. Krämer PM, Baumann RA, Stoks PG (1997) Anal Chim Acta, submitted
65. Pichon V, Chen L, Hennion MC (1995) Anal Chim Acta 311:429
66. Pichon V, Chen L, Hennion MC, Daniel A, Martel F, Le Goffic, Abian J, Barcélo D (1995) Anal Chem 67:2451
67. Pichon V, Chen L, Durand N, Le Goffic F, Hennion MC (1995) J Chromatogr A 725:107

Algal Toxins and Human Health

Ian R. Falconer

Department of Clinical and Experimental Pharmacology. University of Adelaide, Adelaide, South Australia 5005, Australia
E-mail: *ifalconer@medicine.adelaide.edu.au*

The blue-green algae, more correctly called cyanobacteria, are becoming an increasing problem in fresh, brackish and marine waters. They respond to increased nutrient concentrations in the water, particularly soluble phosphorus, to form water-blooms. Frequently these water-blooms are toxic to livestock, and present a potential hazard to human health. Cyanobacterial neurotoxins include anatoxin-a, a neuromuscular blocking agent; anatoxin-a(s) an antiacetylcholinesterase, and saxitoxins, the paralytic shellfish poisons which block axonal sodium channels. Hepatotoxins include the cyclic peptide toxins microcystin and nodularin, which cause severe liver injury and promote tumour growth and the alkaloid toxin cylindrospermopsin which causes widespread tissue injury, including damage to the gastrointestinal lining, liver and kidneys. Exposure to these toxins can be through recreational activities in lakes and rivers, through the drinking water supply, or through water used in renal dialysis. Examples of human illness from these three sources occur. Water treatment can be developed to extract or destroy the toxins. Human epidemiological studies are needed to support the experimental animal studies in order to refine recommendations for guideline concentrations for cyanobacterial toxins in drinking water.

Keywords: blue-green algae, cynobacteria, toxicity tumours, drinking water, health.

Contents

The Handbook of Environmental Chemistry Vol. 5 Part C
Quality and Treatment of Drinking Water II (ed. by J. Hrubec)
© Springer-Verlag Berlin Heidelberg 1998

1
Introduction – The Background to Cyanobacterial Poisoning

The toxic blue-green algae, which are more correctly named cyanobacteria, are procaryotes unrelated to the eucaryotic green algae which are common in freshwater. The cyanobacteria form colonies of photosynthetic cells, either in spongy clusters or connected into filaments [1]. They are very widely distributed, in fresh and marine waters, in soil and on moist surfaces. Mixed species of cyanobacteria occur in all freshwaters, including water supply reservoirs. Some are benthic – that is they adhere to the lake bottom, others planktonic and are suspended in the water as freely floating colonies.

Cyanobacteria only come to the attention of water supply authorities when cell density in the water rises to high concentrations [2]. If the water is visibly coloured by cyanobacteria the cell concentration is likely to be greater than 10,000 cells ml^{-1}, and the occurrence is called a water-bloom. Several species of cyanobacteria contain gas vacuoles within the cells, which rise to the water surface under calm wind conditions. These floating scums may contain 10^6 cells ml^{-1}.

Unfortunately for both agricultural water users, and drinking water suppliers, surface scums of cyanobacteria are often poisonous [3, 4]. The first scientific report of toxicity due to a water bloom was that of Francis in 1878 [5]. The organism was *Nodularia spumigena*, and was described as "forming a thick scum like green oil paint, some two to six inches thick, and as thick and as pasty as porridge". Francis describes it as causing the deaths of sheep, horses, dogs and pigs. Since that time many similar reports have been made, with livestock and pet deaths due to toxic cyanobacteria widely distributed in the Northern and Southern hemispheres [6].

Human injury also occurs as a result of consuming cyanobacterial toxins. Some of the poisoning events were consequences of recreational use of lakes [7], others due to the water reticulation system containing likely cyanobacterial toxins [8–10]. As more toxic cyanobacteria are identified, and the toxins characterised, the public health aspects of cyanobacterial contamination of water supplies become increasingly significant.

The problem of cyanobacterial blooms appears to be worsening, as a result of the rise in nutrient availability in fresh and marine waters. Domestic and industrial detergents, human and animal excreta, and agricultural fertilisers contribute to the nutrient concentrations in water. In seasonally variable rivers, sewage outfalls may form the largest part of the water flow during periods of dry weather. Availability of soluble phosphorus in water appears to be directly related to cyanobacterial blooms, and toxicity [11–13].

Water treatment for the removal of intact cyanobacteria, and dissolved cyanobacterial toxins, has been installed in many countries. The most widely used technique for soluble toxin removal is by activated carbon [14]. Newer approaches are being tested, employing ozone or ultraviolet light [15–17]. For water supply authorities to effectively employ toxin removal techniques, methods for toxin measurement in water must be available. This has proved an ongoing problem, as risk-assessment calculations indicate approximately 1 µg l^{-1} of the toxin microcystin in drinking water as an advisory maximum level and there are no standardised assays presently available [18].

Microcystin, a cyclic heptapeptide toxin, is regarded as the most significant potential source of human injury from cyanobacteria, being both a liver poison and tumour promoter [19]. It is therefore necessary to measure microcystin in drinking water, over the range 0.1–10.0 µg l^{-1}, for water monitoring purposes.

Two types of assay are under development for monitoring, one based on an enzyme-linked immuno-assay technique [20, 21], and the other on the inhibition of specific phosphatase enzymes by microcystin [22, 23]. Both techniques have suitable sensitivities, and are expected to be commercially available in due course.

To link together the monitoring of toxicity of water supplies with measurement of human health is a major step, requiring large-scale epidemiological investigation. A beginning occurred in 1983 with the retrospective evaluation of human liver injury in association with a cyanobacterial bloom on the water supply [24]. Recent prospective surveys of gastroenteritis and skin irritation in association with cyanobacterial counts in the water source showed statistical correlation [25]. In neither of these studies was it possible to measure cyanobacterial toxins in the water supply, due to lack of appropriate methods. In Southern China, epidemiological evidence has linked primary hepatocellular carcinoma with cyanobacterially contaminated surface water supplies [27].

Definitive proof of significant human injury at the level of population health awaits larger scale epidemiological investigation. The present evidence, however, points towards the potential for toxic cyanobacteria in drinking water supplies to pose a real hazard to consumers.

2
Toxic Cyanobacteria (Blue-Green Algae)

The large number of known species of cyanobacteria are classified at present on the basis of cell and filament morphology, and habitat. DNA-based species classification will, in the future, be used to revise the present system. This approach will not only assist in classification, but also in characterisation of toxic and non-toxic strains [28, 29]. Toxicity does not appear to be related to any particular morphological grouping or genus, but occurs in species classified into at least 18 genera. These include the sponge-like cell assemblies of the genus *Microcystis*, as well as the straight filaments of *Cylindrospermopsis* and tight spirals of some *Anabaena* species [30, 31].

Table 1. Toxic cyanobacteria

Anabaena affinis Lemm.	*Lyngbya birgei* G.M. Smith (Thur.)
Anabaena baltica J. Schm.	*Lyngbya major* Menegh.
Anabaena circinalis Rabenh.	*Lyngbya majuscula* Harvey
Anabaena flos-aquae	*Microcoleus lyngbyaceus* (Kutz.) Crouan
(Lyngb.) Breb.	*Microcystis aeruginosa* Kutz.
Anabaena hassallii (Kutz) Wittr.	*Microcystis (Aphanocapsa) farlowiana*
Anabaena lemmermanni P Richt	Drouet et Daily
Anabaena spiroides Kleb.	*Microcystis flos-aquae* (Wittr.) Kirchn.
Anabaena torulosa (Carm.) Lagerh.	*Microcystis toxica* Stephens
Anabaena variabilis Kutz.	*Microcystis viridis* (A.Br.) Lemm.
Aphanizomenon flos-aquae (L.)Ralfs	*Nodularia spumigena* Mertens
Coelosphaerium kutzingianum Nag.	*Nostoc paludosum* Kutz.
Coelosphaerium naegelianum Ung	*Nostoc rivulare* Kutz.
Cylindrospermopsis raciborskii	*Oscillatoria agardhii* Gom.
(Wolosz.) Seenaya et Subba Raju	*Oscillatoria lacustris* (Kleb) Geitler.
Fischerella epiphytica Ghose	*Schizothrix calcicola* (Ag.) Gom.
Fischerella musciola (Thur.) Gam.	*Scytonema hofmanni* Ag.
Gloeotrichia echinulata	*Scytonema pseudohofmanni* Bharadw.
(J.E. Smith) P. Richter	*Spirulina subsalsa* Oerst.
Gloeotrichia pisum Thur.	*Symplaca hydroides* Kutz.
Gomphosphaeria aponina Kutz.	*Tolypothrix conglutinata* var. Colorata Ghose
Hapalosiphon fontinalis (Ag.) Bor. Gam.	*Trichodesmium erythraeum* Ehrenb.
	Umezakia natans gen et sp nov Watanabe [30]

Table 1 lists the presently known toxic species of cyanobacteria in fresh and marine waters, but the list is steadily growing and requires continual revision. Since water and health authorities become concerned when potentially poisonous cyanobacterial numbers reach 'water-bloom' status, it is possible to provide extensive data on the toxicity of water-blooms, allied to the species causing the bloom. In the USA [32], Scandinavia [33], and the UK [34], about 40–75% of all water-blooms tested showed toxicity by mouse bioassay.

A complex and presently unresolved problem occurs with respect to the presence of apparently non-toxic strains of toxic species. In Australia, Baker and Humpage reported that water-blooms in the Murray-Darling Basin showed toxicity in 46% of samples of blooms of *Anabaena circinalis*, 33% of blooms of *Microcystis aeruginosa* (f. aeruginosa), and 80% of blooms of *Nodularia spumigena* [35]. Thus the assumption of water authorities has to be that all water-blooms of cyanobacteria are toxic, unless tested and found non-toxic. Even the evidence of non-toxicity at one site and one date may be deceptive, as separate sampling sites on a single lake often vary in toxicity, and repeated sampling at a single site may show variation in toxicity in consecutive samples [36].

A further source of complexity is the diverse chemical nature of the toxins identified, and their range of mechanisms of toxicity. Some cyanobacterial genera appear to produce only one family of toxins, for example the five identified species of *Microcystis* all produce cyclic heptapeptides of the microcystin family. *Anabaena* genus, by contrast, appears to be able to produce three quite different types of neurotoxins as well as the cyclic heptapeptides [6, 35].

The detailed relationship between toxigenic potential and actual toxicity awaits the application of new, highly sensitive methods of toxin measurement. Identification of genetic elements coding for toxin biosynthesis will greatly assist this process [37, 38]. At present the majority of data for water-bloom toxicity has come from routine mouse bioassay by water supply or health authorities, which will not detect low concentrations of toxins.

Ecological studies with toxic strains of cyanobacteria in culture have not proved to be very informative with respect to factors favouring or opposing toxin production under natural conditions. Water temperature and light intensity both affect toxin production in *Microcystis* and *Oscillatoria*, with the general conclusion that positive stimulation of growth of the cells will also increase toxin production [39–42]. Many of the toxic cyanobacterial species have gas vacuoles, which enable them to move towards or away from the water surface to optimise their conditions for nutrients and light [43, 44]. Thus under field conditions, cell numbers, rather than prior environmental conditions, are used as the best practical guide to the likely magnitude of toxicity of a volume of water.

3
Cyanobacterial Toxins

Cases of poisoning of livestock due to consumption of cyanobacteria are widespread, and have lead to experimental investigation of the nature of the toxins and mechanism of toxicity. Field and laboratory observation of poisoned animals has resulted in the identification of two general areas of toxic effect, which are quite different in clinical symptoms and post-mortem pathology. The most rapidly lethal toxic cyanobacterial extract will kill mice within 10 min of intraperitoneal injection, by terminal respiratory failure. No post-mortem pathological changes to organs can be observed. This outcome is the consequence of neurotoxins which are water soluble and of low molecular weight, rapidly being absorbed into the bloodstream and interfering with nerve function [6].

The other, slower, cause of death from cyanobacterial poisoning will result in death from 20 min to 7 days after injection in mice as a result of liver injury. External symptoms after intraperitoneal injection in mice are progressive loss of pink colour in the ears and eyes, and gasping breaths. In cases of death within 24 h the liver is swollen, dark red and shows extensive internal haemorrhage [45]. After 3–7 days survival, the liver becomes paler and histological examination shows extensive cell death and functional failure.

These two different types of toxicity will be considered separately, together with the chemical nature of the toxins.

3.1
Neurotoxins from Cyanobacteria

A number of genera of cyanobacteria produce neurotoxins, including *Anabaena, Aphanizomenon* [46], *Nostoc* [47], *Cylindrospermum* and *Oscillatoria* [48–50] in freshwater, and *Trichodesmium* [51], in tropical marine waters.

Almost all these identifications of neurotoxicity arose from the investigation of deaths of livestock or domestic animals, poisoned by drinking cyanobacterial scums or eating crusts of drying cyanobacteria on lake shores.

Human illness or poisoning from neurotoxic cyanobacteria has been limited to recreational exposure, rather than consumption of drinking water containing neurotoxins. Though not well characterised, respiratory distress following inhalation of spray containing neurotoxic cyanobacteria has been reported both from marine [51] and freshwater locations [52]. In the latter case a water-skier using a stretch of the River Murray which contained a potentially neuro-toxic *Anabaena* bloom had considerable difficulty in breathing. It was not clarified whether this was an allergic response, or due to neurotoxicity [52].

3.1.1
Anatoxin-a

This was the first neurotoxic alkaloid to be identified, and it was isolated from a culture of *Anabaena flos-aquae* [53, 54]. Structurally the molecule is 2-acetyl-9-aza-bicycle [4.2.1] non-2-ene, with a molecular weight of 65 daltons (Molecule 1) [46].

(Molecule 1)

Anatoxin-a hydrochloride

Analysis for anatoxin-a has employed a variety of methods. The most successful are based on high performance liquid chromatography (HPLC) [55], gas chromatography (GC) [56, 57], and thin layer chromatography (TLC) [58].

The molecule is synthesised in Anabaena from ornithine, by ornithine decarboxylase producing putrescine as the first step [59].

Pharmacologically the action is well characterised, the alkaloid being a powerful depolarising agent at the neuromuscular junction. The action is similar to the effect of acetylcholine and acts at both nicotinic and muscarinic receptors [60, 61, 62]. Because the molecule is not easily cleared or displaced from tissue, the neuromuscular blockade is very sustained making both appli-cation of an antidote and use of artificial respiration impractical [62].

The LD_{50} for pure anatoxin-a is approximately 200 µg kg^{-1} bodyweight for mice by intraperitoneal injection, while the oral LD_{50} for sonicated cell suspen-sions of anatoxin-a-containing *Anabaena* cells is about one hundred to one thousand times higher [63]. The majority of occurrences of anatoxin-a poison-ing have come from Anabaena, and the species concerned were identified as *A. flos-aquae*, *A. spiroides* and *A. circinalis* [63], all planktonic spiral filamen-tous forms. *Oscillatoria* is also a source of anatoxin-a poisoning, in this case benthic growth exposed by decreased water level in a lake [50]. A closely related poison is the propionyl form of anatoxin-a, termed homo-anatoxin-a (2), isolated from benthic *Oscillatoria formosa* (Molecule 2) [49].

The clinical symptoms of poisoning in mice by both anatoxin-a's are identical, with a very short latent period of 1–2 min, then laboured breathing and progressive limb paralysis, abdominal breathing (gasping) violent convulsions or jumping and death by respiratory arrest within 15 min [49]. Similar symptoms – that is staggering, muscle twitching, convulsions, exaggerated abdominal breathing and death from respiratory failure have also been seen in poisoning of large animals [63].

$$H_3\overset{+}{N}C\overset{-}{l}\quad \overset{O}{\underset{9}{\|}}\overset{}{C}-CH_2CH_3$$

(Molecule 2)

Homoanatoxin-a hydrochloride

While about 1–2 g dry weight of neurotoxic *Anabaena* may be lethally poisonous if swallowed by a child, this is unlikely to occur through a drinking water supply. The main reason is that water heavily contaminated with *Anabaena* cell contents has a powerful and unpleasant odour and taste, reminiscent of rotting potatoes. The odour compounds include geosmin and β-cyclocitral, and result in rejection of the water for drinking [64]. Accidental swallowing or inhalation during water sports is possible, and authorities and councils responsible for recreational lakes and rivers widely employ signs (Fig. 1) to warn of poisonous cyanobacteria.

Fig. 1. Warning sign erected on the bank of a water supply reservoir subject to repeated blooms of *Anabaena*

3.1.2
Anatoxin-a (s)

This neurotoxic compound was also isolated from *Anabaena* collected from natural blooms [65]. Due to excessive salivation when administered to laboratory mice, it was named anatoxin-a (s) for salivation. The structure was determined to be a *N*-hydroxyguanidine methyl phosphate ester, of molecular weight 252 daltons (Molecule 3) [66].

(Molecule 3)

Anatoxin-a(s)

Assay techniques for this toxin currently depend on mouse bioassay, or the use of an erythrocyte acetycholinesterase inhibition test. The general process of mouse bioassay for cyanobacterial toxins has been described [67] and the enzyme inhibition test is derived from the methodology for organophosphorus insecticide poisoning [68].

The biosynthetic pathway for anatoxin-a(s) is from L-arginine, with methyl groups from tetrahydrofolate [69]. The pharmacology of this poisoning is well understood, as a consequence of the close parallel between the action of organophosphorus insecticides as inhibitors of synaptic acetylcholinesterase, and anatoxin-a(s). The cleavage of acetylcholine at a very rapid rate is an essential component of nerve impulse transmission across synaptic junctions. The nerve synapse acetylcholinesterase is irreversibly inhibited by synthetic organic phosphate esters such as diisopropyl-fluorophosphate, or by anatoxin-a(s) [63], resulting in the hyperactivity of post-synaptic neurones. The toxicity may be opposed by atropine, as is the case with organophosphorus insecticide poisoning [70].

The LD_{50} for mouse by intraperitoneal injection of pure anatoxin-a(s) is about 20 µg kg^{-1}, which is appreciably more toxic than anatoxin-a. Survival time for a lethal dose is 10–30 min [63]. The symptoms of poisoning in mice and rats include viscous salivation, lacrymation (chromodacryorrhea-bloody tears in rats) urination, defecation, twitching, convulsions and death by respiratory arrest. In large animals, field cases of poisoning from anatoxin-a(s) in *Anabaena flos-aquae* blooms, and experimental studies by dosing, have shown the symptoms described for mice, plus ataxia, dyspnoea, recumbency and cyanosis [70, 71].

The toxin is relatively unstable at elevated temperatures ($> 40\,°C$) and alkaline pH. It is also of low toxicity by mouth to rodents [71]. However, it has high oral toxicity to pigs, which have a digestive system comparable to ourselves, and hence represents a potential risk to human consumers. However, no reports exist of human injury from water supplies or recreational exposure to *Anabaena* containing this neurotoxin.

3.1.3
Saxitoxin and Related Neurotoxins

The saxitoxins are well known as the paralytic shellfish poisons associated with marine dinoflagellate 'red tides' [72]. The written history of paralytic shellfish poisoning goes back to 1793, when men from two boat crews on the exploration voyage of Captain George Vancouver to the coast of British Columbia were affected. Four men who had eaten roasted mussels were ill and one died [72].

Only recently, however, has the significance of these poisons as cyanobacterial neurotoxins been clarified. A water bloom of *Aphanizomenon flos-aquae* in New Hampshire, USA, was associated with a large fish-kill when the local authorities treated the bloom with copper sulphate [73]. The electrophysiological properties of the toxins (called aphantoxins) were similar to the paralytic shellfish poisons [73, 74]. These toxins were later identified as saxitoxin and related neurotoxins [75, 76]. Only in New Hampshire has *Aphanizomenon* been reported with these toxins.

In Australia, livestock (sheep) deaths caused by neurotoxicity after consuming *Anabaena circinalis* in farm water supply dams resulted in laboratory investigations, without identification of the nature of the toxin [77, 78]. The extensive *Anabaena* water bloom covering 1,000 km of the Darling River system in 1991 caused thousands of sheep and cattle deaths, due to the highly neurotoxic cyanobacterium [79]. Further investigation of neurotoxic *Anabaena* from widely separated locations on the Darling River, sites on the River Murray, and bloom samples from six water supply reservoirs, all showed the presence of paralytic shellfish poisons. The most toxic of these water blooms, from a domestic drinking water supply reservoir, had a minimum lethal dose for mice by intraperitoneal injection of $8-16$ mg dryweight of *Anabaena* kg^{-1} bodyweight, comparable in neurotoxicity to earlier samples lethal to sheep [78].

The paralytic shellfish poisons present included saxitoxin, neosaxitoxin and gonyautoxins 1, 2 (the most abundant) 3, 4, 6, and decarbamoyl 2 and 3 [80]. These molecules are closely related tricyclic guanidinium alkaloids [4] with molecular weights largely between 300 and 400 daltons (Molecule 4) [80].

(Molecule 4)

R = H; saxitoxin dihydrochloride
R = OH; neosaxitoxin dihydrochloride

Analysis of paralytic shellfish poisons is made more complex by the large number of related molecules, and mouse bioassay remains the official standard [81]. A progressive development of HPLC techniques is, however, providing accurate quantitation of the range of related toxins [80, 81].

The pharmacology of the paralytic shellfish poisons has been extensively studied, partly as a result of the fundamental role of ion channels in axonal nerve impulse transmission. The toxins cause a relaxant action on vascular smooth muscle, a depression of the action potential of cardiac muscle, and inhibition of axonal impulse transmission through blocking the voltage-regulated sodium channel [72, 82]. The guanidinium 7, 8, 9, ring carrying a positive charge is essential for the blocking action, and the depression of sodium entry is dose dependant [82].

There is no pharmacological antidote to paralytic shellfish toxin poisoning, though dosing with activated carbon is likely to be of benefit through toxin adsorption [72]. Inducing vomiting is commonly used in shellfish poisoning of any type. The toxicity of pure saxitoxin is 10 μg kg^{-1} bodyweight (intraperitoneal in mice), but other related molecules show a range of largely lower toxicities [80, 83].

The clinical symptoms of poisoning in farm animals are incoordination followed by recumbency and death by respiratory failure [78]. In man the initial symptoms are rapid, from 30 min to 3.3 h after ingestion of toxic shellfish, with paraesthesia and numbness of lips and mouth, extending to the face, neck and limb extremities. Motor weakness follows with incoordination, and in severe cases respiratory difficulty and muscular paralysis [72].

The implications of paralytic shellfish poisons for human health arise from three possible sources of ingestion. In the case of *Anabaena circinalis* of high toxicity in drinking water supply reservoirs, the possibility exists for poisoning if a dense water bloom is treated with copper sulphate, releasing the soluble toxins into the water. This is the analogous situation to the observed fish-kill [73]. A second possibility is from recreational water use, which can lead to swallowing poisonous scum while swimming, windsurfing, water-skiing, or simply falling out of a boat. The third and obvious source of toxicity is from consumption of freshwater molluscs growing in a neurotoxic bloom, which have now been shown to accumulate paralytic shellfish poisons from *Anabaena* [84]. To date, no proven instances of human injury from paralytic shellfish poisons in freshwater cyanobacterial blooms have been reported, the only possible case being a water-skier with respiratory difficulties after skiing through an *Anabaena* bloom [52].

3.2
Hepatotoxins from Cyanobacteria

The majority of reported cases of human illness, and of livestock deaths from cyanobacterial toxicity, have been brought about by organisms producing hepatotoxins. This is a consequence of the widespread distribution of these toxic organisms and their ability to form water blooms in diverse habitats.

The most abundant genus producing hepatotoxins is *Microcystis*, which is a common bloom-forming species in lakes, reservoirs and farm water storages from the tropics to cold-temperate regions in both hemispheres. It is capable of forming massive water blooms in nutrient-enriched lakes, resulting in deaths of livestock. Examples have been reported from South Africa, North America and Australia [85–87].

Toxic compounds from this species have been named microcystins, and form a family of closely related cyclic heptapeptides [88]. These are, however, not restricted to the genus *Microcystis*, and also occur in quite unrelated cyanobacterial genera.

Brackish water in coastal lakes and lagoons also provides a habitat for toxic cyanobacterial blooms, as does the Baltic Sea. The species concerned is *Nodularia spumigena*, first described by Francis in 1878 [5]. The toxin in this case is a cyclic pentapeptide, named nodularin [89]. Almost all *Nodularia* samples tested have mouse toxicity [28].

Toxic cyanobacteria in tropical waters have been less studied, possibly as a consequence of more urgent health problems from pathogens and parasites in tropical water supplies. However, a common cyanobacterium in tropical lakes and water storages *Cylindrospermopsis raciborskii* contains a powerful non-selective tissue toxin which has caused severe human illness [10]. It is likely that cases of poisoning by this toxin will be increasingly reported, now that it has been clearly identified.

Systematic monitoring of cyanobacterial hepatotoxins in water is beginning in several countries, on a project basis. For example, it was reported in 1996 that Dr. W. W. Carmichael would be heading a two year study of raw (natural), treated and 'during treatment' drinking water in 20–30 water utilities in the USA in 1996–97 [90]. Similarly projects are at the design stage for pilot epidemiological evaluation of human injury associated with hepatotoxic cyanobacteria in water supplies [91].

3.2.1
Microcystins

These toxic peptides are characteristically found in several species of the genus *Microcystis*, including *Microcystis aeruginosa, M. viridis* and *M. wesenbergii*, all colonial aggregates of spherical cells in a mucilaginous sponge-like structure [92]. They also occur in a large number of diverse species, sometimes in conjunction with neurotoxins, for example in *Anabaena* and *Nostoc*. The majority of the species containing microcystins are filamentous organisms which are found in a very wide range of habitats [93].

The original identification of the structure of one of the most abundant molecular forms of microcystin (microcystin-LR) was carried out by Fast Atom Bombardment/Mass Spectrometry at the University of Cambridge in 1984, with pure microcystin from a South African water bloom [94]. Prior to this the basic amino acid composition of another microcystin had been identified, but the unique hydrophobic amino acid ADDA (3-amino-9-methoxy-10-phenyl-2, 6, 8 trimethyl deca, 4, 6 dienoic acid) was not observed [95]. The location of L-amino acids within the microcystin family of molecules became clear with the publication of the structures of four more microcystins, which each contained two variable L-amino acids, the five variants being – leucine, alanine (LA); leucine, arginine (LR); tyrosine, arginine (YR); tyrosine, alanine (YA); tyrosine, methionine (YM) (Molecule 5) [96].

(Molecule 5)

Microcystin

Since that time over 50 variants of microcystins have been reported, and the essential nature of the ADDA residue for toxicity identified [97–99]. Molecular weights of microcystins range from 909–1067 daltons [99]. The structure of ADDA has been confirmed by total synthesis [100]. It should be noted that one consequence of the variable L-amino acids is the marked difference in lipophilic/hydrophilic character between the leucine, alanine or tyrosine, methionine forms and the arginine, arginine form of the molecule.

The biosynthetic pathway for microcystin has been studied; ADDA appears to be derived from phenylalanine and acetate and the intermediate products are linear peptides, containing both L- and D-amino acids [101, 102].

The presence of D-amino acids in the peptide implies that non-ribosomal peptide synthetase enzymes are likely to be involved. A recent study of the genomic DNA in several *Microcystis* strains of differing toxicity, using hybridisation with fragments of core sequences from bacterial and fungal peptide synthetases, identified extensive homology. Comparison of the *Microcystis* DNA nucleotide sequences in the hybridising regions, with known sequences from peptide synthetases from other species, showed closest homology with non-epimerising enzymes, which is consistent with non-ribosomal synthesis from D-amino acids. A DNA fragment from a toxic *Microcystis* strain potentially coding for a complete unit of a peptide synthetase enzyme was observed to hybridise with DNA from all toxic strains, but not with DNA tested from non-toxic strains. Thus a real genetic difference is inferred between these toxic and non-toxic strains of *Microcystis* [38].

Analysis of the microcystins has received very considerable attention, due to the importance for public health of measuring microcystins in water supplies and recreational waters. For research purposes, the main technique is HPLC analysis for purification, separation and quantitation of microcystins. The low sensitivity of this method requires one or more prior concentration steps, from intact cyanobacteria or from water samples [99]. Use of solid-phase C-18 adsorbtion cartridges for concentrating microcystins from treated drinking water or relatively clean natural waters has been widely and successfully employed [99, 103]. Structural identification of particular microcystins has used Fast

Atom Bombardment Mass Spectroscopy and Nuclear Magnetic Resonance techniques [99, 104].

Water supply monitoring requires detection sensitivity in the low ng ml^{-1} range, and ability to handle many samples at moderate cost. The appropriate techniques were developed using two independent methodologies, Enzyme Linked Immunosorbent Assay (ELISA) and enzyme-inhibition assay. The immunological methods depend on the specificity of antibodies raised against toxin-protein conjugates. These are then used as antigens for raising serum antibodies in rabbits, or for monoclonal antibody production from mouse lymphocytes. The polyclonal rabbit serum antibodies have been successfully developed into standardised assays [105 – 107]. It is expected that a commercial kit based on these assays will be available in 1997.

An alternative mechanism for antibody production used inoculated chickens, which secrete antibodies into the egg yolk. An immunoassay based on this technique has recently been used to measure ng ml^{-1} concentrations of dissolved microcystins in river and lake waters in northern USA. The highest concentration of free microcystin measured was 200 µg l^{-1}, with several samples from widely separated locations containing 10 – 60 µg l^{-1}. *Microcystis* cells showed a range of toxin concentrations from 1.9 mg g^{-1} to 12.8 mg g^{-1}, which is in general agreement with other data [108].

The alternative method for measurement of nmolar concentrations of microcystins is to use the biological action of the toxin. The microcystins exert their toxic effect through inhibition of specific intracellular phosphatases (1 and 2A) [109 – 111]. The inhibition constants are very low, 1 C_{50} for Type 1 protein phosphatase = 1.7 nM, 1 C_{50} for Type 2A protein phosphatase = 0.04 nM, both with microcystin-LR [112]. Recent development of colorimetric assays for inhibition of protein phosphatase by microcystins provides a direct low-cost technique applicable to water testing [107, 113]. The key to these methods is recombinant protein phosphatase-1 as the inhibited enzyme, and the synthetic substrate paranitrophenylphosphate as colorimetric substrate [113]. This avoids use of ^{32}P-labelled protein substrates [114, 115] and difficult enzyme purification procedures.

Radioactive ^{32}P substrates are, however, of value for laboratory investigation, and have been applied with success to large scale screening of cyanobacterial cultures for microcystins and related toxins [93].

3.2.1.1
Mammalian Toxicity

Oral ingestion, inhalation or experimental injection of microcystin-containing cyanobacterial cells or extracts will cause the death of mammals. Small mammals such as mice die from acute poisoning often within 20 m following lethal intraperitoneal injection. The clinical symptoms are pallor and panting, and the cause of acute death is blood loss into the liver which is seen to be swollen and dark red at post-mortem [45]. Large domestic animals survive longer after oral dosing, lethally poisoned sheep dying between 18 – 48 h [116]. In all cases extensive liver damage is observed, hepatocyte necrosis often commencing around the centrilobular veins and extending into almost total cell death [116].

For oral poisoning, the toxin must pass from the gastro-intestinal tract into the circulation in order to reach the liver. Microcystins are water-soluble to varying degrees and do not readily enter most cells and organs [117–119]. Transport from the intestine is via the absorbtive epithelium, utilising in all probability the bile acid transporter system [120, 121]. Entry into hepatocytes also uses the bile acid transporter or a closely related anion carrier, hence resulting in active concentration of microcystins into these cells from the blood [122–124]. Experiments on the uptake of microcystin into hepatocytes in suspension showed a concentration ratio in hepatocytes of 70:1 after 1 h. This uptake was inhibited 70–80% by the bile salt sodium deoxycholate (1 μM) [124].

Initial microcystin uptake into the liver is rapid, and cell changes that can be observed histologically, occur within 15 min of an intraperitoneal dose. In most *in vivo* experiments microcystin-LR has been used, and the initial cell injury has been observed to be in the central regions of the liver lobule. With increasing time the damage extends back towards the portal blood vessels until effectively all the tissue is showing disorganisation, cell swelling and lysis and accumulation of erythrocytes [125]. By contrast, injection of microcystin-YM, which is a more lipophilic molecule, causes initial damage to perilobular hepatocytes, progressing to mid and centrilobular damage [45].

The nature of the hepatocyte injury is best seen in preparations of isolated hepatocytes which can be observed by electron microscopy. In suspension the characteristic spherical hepatocyte with a uniform covering of microvilli is rapidly deformed in the presence of microcystin into a blebbed, smooth surfaced cell [126].

Examination of the mechanism underlying the cell deformation has shown dramatic changes in the cell cytoskeleton. The microfilaments detach from anchoring points at the cell periphery and contract into one or more fibrous masses, the intermediate filaments which form the main structural elements of the cell cytoskeleton disorganise and withdraw to a zone adjacent to the nucleus, and the microtubules dissociate [127–129]. These structural changes account for the cell deformation observed in suspensions of hepatocytes, and the loss of lobular architecture in the liver after poisoning.

The biochemical basis for these changes is beginning to be understood, as a result of the research on phosphatase inhibition by microcystin referred to earlier [109–111]. The aggregation of cytokeratins into intermediate filaments appears to be directly related to the phosphorylation of the cytokeratins, the increase in phosphorylation leading to disaggregation of the filaments [128–131].

The diarrhetic shellfish poison okadaic acid also acts through inhibition of phosphatases 1 and 2A, in a manner parallel to microcystin [132]. As it is a lipophilic compound which readily penetrates into a range of cell types in vivo and in vitro, it has been more widely investigated. In addition to causing cell injury okadaic acid has also been identified as a tumour promoter, as is discussed later.

Experimental liver injury caused by microcystins has thus been well investigated, but field cases of poisoning in animals and people have shown a wider range of symptoms. Some are probably secondary to liver damage, for example photosensitivity and pulmonary emboli [133, 134]. Others are as yet unexplained, though often observed, such as petechiae and ecchymoses haemorrhages in a range of locations including subcutaneously and on the intestines [116].

Chronic and sub-chronic exposures to microcystins in drinking water have shown progressive liver injury with hepatocyte death, white cell invasion of injured tissues, and fibrosis [135]. Mortality in male mice exceeded that in female mice, and deaths during chronic exposure to low concentrations of microcystins were largely caused by bronchopneumonia [135]. Changes in blood characteristics related to liver dysfunction were shown in mice [135] and in a sub-chronic oral dosing of pigs with three dilutions of a toxic *Microcystis* extract [18]. Changes in plasma alkaline phosphatase, γ-glutamyl transferase, total bilirubin and albumin were monitored in 4 groups of pigs receiving 1.3 mg kg^{-1} day^{-1}; 0.8 mg kg^{-1} day^{-1}; 0.28 mg kg^{-1} day^{-1}; 0 total microcystins. Progressive increases were observed in γ-glutamyl transferase and total bilirubin in the two higher dosed groups, and decreases in plasma albumin. Alkaline phosphatase showed an abrupt rise after the start of dosing in the highest dose group, which fell back to normal by 30 days. Aspartate aminotransferase, alanine aminotransferase, and lactate dehydrogenase in plasma did not show clear dose-related changes [18].

3.2.1.2
Human Injury

Examination of plasma indicators of liver dysfunction has been used in an epidemiological study of human injury. A water supply reservoir providing a single town with drinking water was identified as having recurrent blooms of toxic *Microcystis*. The water treatment plant was separated from the supply reservoir by approximately 20 km of pipeline, and an altitude change of 150 m. Under such conditions, including pressure reduction valves in the pipeline, cyanobacteria can be expected to be lysed prior to entering water treatment. The treatment plant was a standard alum flocculation, soda ash pH adjustment, tank sedimentation and rapid sand filtration unit, with chlorine added initially with the flocculant and ending with post-chlorination to ensure residual chlorine in the distribution system. Fluoride was also added.

A toxic bloom of *Microcystis* was monitored in 1981 in the supply dam, over the period of bloom growth and after the water authority killed the bloom by aerially-applied copper sulphate.

For the epidemiological analysis, all relevant data from the Regional Pathology Laboratory were sorted into two groups, the groups being data from plasma samples from people drinking the affected supply, and data from those from adjacent towns on other, unaffected supplies. These two groups were then sorted by date of blood sampling into the 5 week period prior to the water bloom, 5 weeks comprising 3 weeks during the bloom and 2 weeks after copper sulphate treatment, and for the 5 weeks after that. Statistical analysis showed a significant rise in plasma γ-glutamyl transferase during the bloom period, compared with before and after the bloom, only in the group drinking affected water. Alanine amino-transferase in plasma showed parallel changes of lower magnitude, whereas as-partate aminotransferase and alkaline phosphatase showed no difference in the people drinking affected water with time [24]. Care was taken to evaluate poten-tially conflicting sources of liver damage, such as alcoholism and hepatitis, and no variation attributable to these factors influenced the results.

A far more severe occurrence of injury from microcystin poisoning occurred in Brazil in February 1996. At a renal dialysis clinic 110 of 131 patients showed acute nausea, vomiting and visual disturbance after routine treatment. By the following month twelve patients had died from acute haemorrhage and seizu-res, by August fifty-five patients had died, all with liver haemorrhage and failure. Examination of the water reservoir, the dialysis centre filter, and the patients blood and liver post-mortem, all demonstrated microcystin. The con-centrations in blood and liver resembled those found in acute toxicity of mice [136]. The histopathological examination of the liver of the patients was con-sistent with microcystin toxicity.

Dialysis with water containing a hepatotoxin is equivalent to an intravenous injection of the toxin and greatly more damaging than an oral dose of the same quantity [67].

The other published medical report of human injury arising from exposure to toxic *Microcystis*, came from accidental „recreational" type exposure of army recruits, who were swimming in full pack, and doing "eskimo rolls" in canoes, in a lake with a heavy *Microcystis* bloom [7]. The two most affected soldiers had pneumonia, as well as blistered mouths, sore throats, vomiting, diarrhoea, head-aches, and abdominal pain. Eight other soldiers who had been canoeing had the latter symptoms without the pneumonia, but all had dry coughs. Plasma from the most affected soldier showed alanine and aspartate aminotransferases were above the normal range.

Toxic water blooms of *Microcystis*, identified as a result of livestock poison-ing, often show lethal toxicity around 30 mg dryweight of cells kg^{-1} bodyweight of mice by intraperitoneal injection. Toxin content of this material is of the order of 2.4 mg toxin g^{-1} dryweight of cyanobacteria. Often multiple variants of microcystin are present, with one predominant [18, 99].

On the basis of cell counting, it was calculated that the toxin content per *Microcystis* cell was approximately 0.2 pg for a bloom of toxicity of 30 mg kg^{-1} bodyweight [18].

3.2.1.3
Tumour Promotion

Indications of the possibility of tumour promotion, but not organ-specific car-cinogenicity, came from a study of chronic oral *Microcystis* consumption in mice [135]. This result was followed by a direct demonstration of skin tumour promotion by oral *Microcystis*, (containing microcystin YM) after initiation by topical application of dimethylbenzanthracene to the skin [137, 138]. In the rat, intraperitoneal injection of microcystin-LR has been shown to increase the foci of potentially hepatocellular carcinoma cells in liver, following initiation by diethyl nitrosamine [139].

The clarification of the mechanism of action of microcystin through inhibition of protein phosphatases, has greatly assisted in interpretation of the tumour promoting affects. The mechanism of action is effectively identical to okadaic acid, which had been demonstrated to be a potent tumour promoter of wider cell specificity than microcystin [132]. Both inhibitors result in a large increase in protein phosphorylation in cells, and associated changes in cell division which may be exhibited as increases in cell proliferation [140, 141].

Tumour promotion was first explored as a result of the demonstration of irritant plant exudates containing phorbol esters, promoting skin tumour growth. The mechanism of action was shown to involve protein kinase C, which phosphorylated a variety of intracellular proteins [142]. Thus the microcystins, okadaic acid and phorbol esters, all tumour promoters, act through increased protein phosphorylation. How this relates to accelerated cell division of potential tumour-forming cells is not clear, but is currently subject to extensive investigation and speculation [132].

The possible organ sites for microcystin-stimulated tumour growth also require investigation, since primary hepatocellular carcinoma is a rare cancer in developed countries. Gastrointestinal tract carcinomas are of greater quantitative importance in Europe, North America and Australia, and are potential locations for microcystin-stimulated carcinogenesis [140, 143].

In underdeveloped areas, where liver cancers are more frequent due to high exposures to hepatitis B, aflatoxins in the diet, and poor nutrition, an association with cyanobacterial contamination of water supply is more likely. In the Shanghai region of China, the epidemiology of hepatocellular carcinoma in local rural areas has been extensively studied. A strong statistical relationship has been shown between the drinking of surface water, and disease incidence. The surface water contains abundant cyanobacteria and measurable microcystin concentrations. Preliminary comparison of the drinking water consumed in an area of high disease incidence by hepatocellular carcinoma cases (0.61 μg l^{-1} microcystins) and controls free of the disease (0.36 μg l^{-1}), showed a clear difference. Deep well water has no detectable microcystins, and is consumed in the villages with negligible incidence of hepatocellular carcinoma [26, 27]. In these areas of high endemic hepatocellular carcinoma, public health policy includes vaccination for hepatitis B, dietary changes to avoid alfatoxins, and provisions of well water [27].

Macrophages appear to be responsive to microcystins, which may have implications for white cell involvement in the progressive liver injury seen on chronic exposure to microcystins, and also for tumour promotion. Cultured mouse peritoneal macrophages respond to microcystin by producing interleukin-1, arachidonic acid release, prostaglandins and tumour necrosis factor-α [143–145]. These compounds are characteristic of inflammatory tissue responses, and may play a part in tumour growth [132].

3.2.1.4
Public Health

From the preceding detailed discussion of the toxic effects of microcystins, and the wide range of cyanobacteria which produce them, it is apparent that their presence in water supplies is an issue relevant to public health. There are two separate issues of concern. One is the acute liver injury occurring after a short exposure to the toxins, which may have a cumulative effect on liver malfunction with pre-existing hepatitis and alcoholism. The other is the injury occurring with chronic low-concentration exposure over long periods. This may be manifest in reduced liver function alone, or by increased rates of hepatocellular carcinoma in addition to reduced function. The possibility of a relationship between gastrointestinal tumours and microcystin contaminated water supply is under investigation [19].

3.2.1.5
Water Guidelines

A crucial question for less developed countries is the relationship between surface water quality and liver cancer. For developed countries the relationship of microcystins in water supplies to liver injury and tumour incidence in the general sense is critical. The formulation of water quality guidelines for microcystins is highly dependent on the resolving of these relationships. Experimental sub-chronic oral exposure to microcystins has been carried out in mice and pigs to provide Lowest Observed Adverse Effect Levels (LOAEL), or No Observed Adverse Effect Levels (NOAEL). The outcome of measurements in both species by independent researchers is similar. Research by the Canadian Environmental Health Directorate lead to the application of an uncertainty factor of 3,000, to a NOAEL of 40 μg kg^{-1} bodyweight day^{-1} determined in mice. From this data was calculated a possible Tolerable Daily Intake of 0.0133 μg kg^{-1} days^{-1}, and a guidance value of 0.5 μg l^{-1} for water supply. Within the 3,000 uncertainty factor was a multiple of 3 for possible tumour promotion [146].

The study carried out in male pigs resulted in a LOAEL of 280 μg kg^{-1} day^{-1}, for 44 days exposure. An uncertainty factor of 1000 was derived from 10 $\times$ to allow for a short sub-chronic LOAEL to be converted to a lifetime NOAEL; 10 $\times$ for use of pig data for human guidelines and a further 10 $\times$ for the susceptibility range within the human population. Application of this to drinking water concentration provides a guidance value for a 60 kg adult drinking 2 l day^{-1} of 8.4 μg l^{-1} of microcystin, or for a 10 kg child drinking 1 l day^{-1} of 2.8 μg l^{-1} [18]. Use of an additional uncertainty factor for tumour promotion of 10 $\times$ gives a range of 0.28–0.84 μg day^{-1} [18]. It is considered by the author that the accumulating evidence for tumour promotion justifies the higher uncertainty factor.

As the epidemiological analysis of microcystin injury to human populations strengthens, it will become necessary to incorporate water guideline values into EEC and WHO recommendations for drinking water.*

---
* See End-note on page 77.

3.2.2
Nodularin and Motoporin

The toxicity of *Nodularia spumigena* to livestock, pets and people exposed to this cyanobacterium is of increasing concern, as the incidence of water blooms of the species grows. It has been extensively reported in the brackish water of river estuaries, and estuarine lakes, from the last century onward [147, 148]. The recent reports of the very extensive occurrence of *Nodularia* in the Baltic Sea, and associated pet deaths from dogs swimming in the sea, has highlighted the issue of cyanobacterial eutrophication in marine environments [149–151]. The very large water blooms of *Nodularia* in the Baltic Sea and consequent ecological problems, require detailed understanding of the bloom dynamics, nutrient inputs and the hydrology of the sea bed [152].

Nodularia blooms are almost all toxic by mouse bioassay, showing some variation in toxicity, but overall to be regarded as highly toxic [153].

(Molecule 6)

The toxic molecule, named nodularin, is a cyclic pentapeptide related to the larger heptapeptide microcystins. Both contain ADDA as an essential amino acid for toxicity [89]. Nodularin comprises a peptide ring containing γ linked D-glutamic, β-linked D-*erythro* methyl aspartic, L-arginine, 2 (methylamino)-2-dehydrobutyric and ADDA (Molecule 6) [89]. Molecular weight is 824 daltons.

The original isolations of nodularin were from bloom and culture material collected as far apart as New Zealand [89] and Finland [154], indicating both the wide distribution of toxic *Nodularia* and the common single structure of the toxin. Analysis of nodularin has been undertaken by HPLC techniques [148, 151, 153], and is also suitable for assay at nanomolar concentrations by the phosphatase-inhibition assays [107]. The available polyvalent rabbit antibodies also cross-react with nodularin to provide a quantitative and sensitive assay [107].

3.2.2.1
Nodularin Toxicity and Tumour Promotion

The acute effects of intraperitoneal injection of extracts of *Nodularia* or purified nodularin in mice are identical with those of *Microcystis* extracts. The initial injury is structural disorganisation of the liver, centrilobular hepatocyte necrosis, and extensive haemorrhage. Isolated hepatocytes show characteristic structural deformation and blebbing. Enzymes indicative of liver injury appear rapidly in the peripheral circulation [155].

The LD_{50} of *Nodularia* blooms can reach 25 mg dryweight of cells kg^{-1} bodyweight of mice, and the toxicity of pure nodularin is approximately 70 µg kg^{-1} by intraperitoneal injection into mice. Thus water blooms can contain about 3 mg of nodularin g^{-1} dryweight of cells.

The mechanism of action at the cellular level has been shown to be identical to microcystin, with a specific inhibitory action on phosphatases 1 and 2 A, of approximately double that of microcystin-LR ($1C_{50}$ microcystin LR = 1.6 nm; nodularin = 0.70 nm [111].

Recent investigation of the tumour-promoting activity of nodularin has shown that it is appreciably more potent than microcystin-LR in stimulating the growth of pre-cancerous nodules in the rat liver [156]. This is in agreement with the greater phosphatase-inhibition activity of the compound, and may also imply greater cell penetration.

3.2.2.2
Public Health

Because *Nodularia* can grow in brackish water, it is most commonly found in estuarine environments. Here it can form a hazard to domestic animals and pets from drinking or coat cleaning, and also a hazard to recreational water users [148]. It is also a source of odour and a general nuisance to water users, sufficient for governments to adopt expensive methods of bloom reduction. In Western Australia a channel through the coastal dunes to the Peel-Harvey Estuary has been cut to provide increased water flow and salinity to minimise *Nodularia* blooms [157, 158].

Drinking water supplies are also drawn from water bodies containing *Nodularia*, and may have to be shut-off when cell counts indicate a potential danger to public health [6]. Filter feeding molluscs can accumulate toxicity from water blooms of *Nodularia*, thus requiring the monitoring of mussel beds in affected areas [159]. Measurement of nodularin in drinking water supplies will be needed in future in locations where water blooms are endemic in the supply reservoir or lake.

Motoporin is a variant on nodularin, having the arginine residue replaced with valine. The compound has been isolated from a tropical sponge species [160]. On the basis of the increased hydrophobicity of this structure, cell penetration may be enhanced, without loss of protein phosphatase inhibition. It has not yet been reported from cyanobacteria.

3.2.3
Cylindrospermopsin

This alkaloidal toxin has now been isolated from two species of cyanobacteria, the common tropical organism *Cylindrospermopsis raciborskii*, and a new tem-

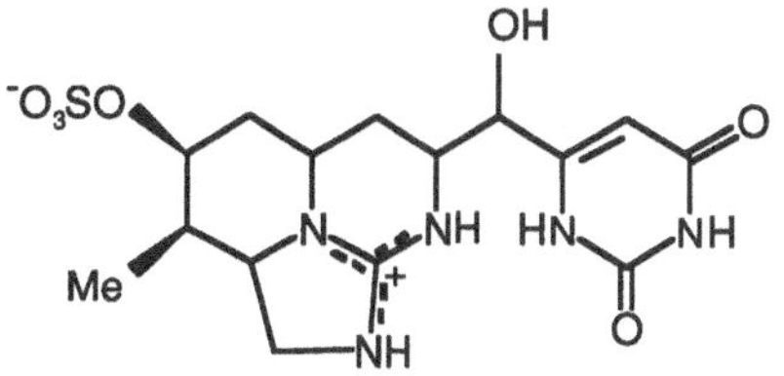

(Molecule 7)

Cylindrospermopsin

perate species from Japan named *Umezaki natans* [161, 162]. The toxin contains a tricyclic guanidine combined with hydroxymethyl uracil (Molecule 7) [161]. Molecular weight is 415 daltons.

Development of an analytical method for cylindrospermopsin has relied on HPLC separation of semi-purified extracts [162].

3.2.3.1
Toxicity

Cylindrospermopsin was isolated as a result of human poisoning attributed to *Cylindrospermopsis* in a water supply reservoir [10]. Experimental dosing of mice lead to widespread tissue injury with particularly acute liver injury [10]. The early centrilobular hepatocyte necrosis is followed by extensive fat storage in the tissues [162], presenting a picture entirely different from *Microcystis* poisoning [45]. The initial toxic action includes the inhibition of protein synthesis, and the proliferation of smooth-surfaced endoplasmic reticulum [163]. Accelerated metabolism of thiols appears to be an important component of toxicity, since cylindrospermopsin sharply lowers the concentration of reduced glutathione and reduces glutathione synthesis in isolated hepatocyte suspensions [164, 165]. Inhibition of cytochrome P450, which is widely involved in oxidative attack on xenobiotics in the liver, provided partial protection of hepatocytes from toxicity, implying that toxic metabolites are involved as well as the parent cylindrospermopsin [165].

It is also possible that toxins other than the identified cylindrospermopsin occur in this organism. Kidney damage was observed in the human poisoning, and is not as evident after administration of the purified toxin. Samples of *Cylindrospermopsis* show variations in toxicity, both in acute death and in the extent of proximal tubule damage that occurs [166].

3.2.3.2
Public Health

Human injury due (in all probability) to *Cylindrospermopsis* poisoning has been documented. A water supply reservoir on tropical Palm Island off the North Queensland coast of Australia provided chlorinated, unfiltered drinking water to an indigenous community. The reservoir had a major cyanobacterial bloom, resulting in complaints about the water taste and smell. The water supply authority dosed the reservoir with copper sulphate killing the bloom. Within one week a severe outbreak of hepatoenteritis occurred, largely in children of average age 8.4 years. No pathogens were detected. About 140 persons were treated, including 85 who were flown to a large regional hospital. Sixty-nine percent of patients required intravenous therapy, largely for severely deranged electrolyte balance [167, 168]. The toxic *Cylindrospermopsis* which was subsequently collected from the reservoir, and grown in culture for investigation, caused extensive kidney damage in mice on intraperitoneal injection, as well as liver and other injury [10].

In the human cases, loss of electrolytes, glucose and protein, and, in some 20% of cases, blood in the urine, showed extensive kidney damage in the poisoned individuals. Together with injury to the intestinal lining leading to profuse bloody diarrhoea, the most acute clinical aspect of these cases was fluid and electrolyte loss. Some patients were so adversely affected that they showed „hypovolaemic/acidotic shock characterised by extreme lethargy" [167].

Since the original poisoning considerable review and medical discussion has taken place on *Cylindrospermopsis* toxicity in Australia, with respect to its role in Palm Island and in a common enteritis in Northern Australia in the past [168–171]. It is likely that toxic *Cylindrospermopsis* is relatively abundant in water reservoirs, rivers and eutrophic lakes in tropical areas worldwide, but that the symptoms from poisoning are merged into a complex health background of parasites, viral hepatitis and enteritis and bacterial pathogens.

In Australia increased awareness of this toxic species has resulted in monitoring of water supplies and toxicity testing of water blooms of *Cylindrospermopsis*. As a result a major supply reservoir was withdrawn from use for about two months in 1995, when routine monitoring followed by mouse bioassay showed the presence of a toxic water bloom of *Cylindrospermopsis* in the reservoir.

4
Skin Irritants and Allergies

One of the commonly observed consequences of swimming in a cyanobacterial bloom is skin irritation [6]. In cases where the cause has been identified, allergic reactions form an important part [172–174]. Respiratory reactions are also common from cyanobacterial allergy, and an extensive study has shown that cyanobacteria may act as inhalant allergens in Type 1 hypersensitivity [175].

However, the toxic effects of cyanobacteria also include skin irritation, which may be so severe as to cause full-thickness skin necrosis [176]. In particular toxins from several marine cyanobacteria including *Lyngbya majuscula*, *Oscillatoria negroviridis* and *Schizothrix calcicola* are skin irritants and tumour promoters, similar to the phorbol esters [177]. Reaction to contact with filaments of *Lyngbya majuscula* as a result of swimming are severe, with a gradual itching and burning sensation, followed by reddening, blistering and deep desquamation. This is particularly painful under the swimsuit, in the areas of the scrotum, perineum and anus [176]. The public health implications of topical exposure to very active tumour promoter compounds are significant [178].

The majority of cases of dermatitis observed after exposure to *Anabaena*, *Aphanizomenon*, *Gloeotrichia*, *Nodularia*, and *Oscillatoria*, have not been specifically attributed to toxicity or allergy [179, 180].

The frequency of skin irritation following exposure to cyanobacteria is of general health concern, largely through recreational or accidental contact. However, in one case of a persistent cyanobacterial bloom on a lake used as a water supply, residents in the town supplied from the lake complained that simple showering caused a skin rash [2, 181]. Long-term effects have yet to be defined.

5
Water Treatment

Water supply authorities have supported both laboratory and pilot plant studies on the removal of cyanobacteria and their toxins, as a result of public complaints of taste and odour, and the potential adverse impacts of these toxins on public health.

Normal water filtration processes employ pre-chlorination as a first step, which lyses cyanobacteria and releases toxins into the water [182]. The subsequent flocculation, rapid sand filtration and post-chlorination do not reliably remove cyanobacterial neuro or hepatotoxins [183–185].

Subsequent research has explored two options, the removal of unlysed cyanobacterial cells, and the adsorbtion or destruction of the toxins. Removal of cells is effectively done by dissolved air flotation of unchlorinated raw water treated with flocculant [186]. Evaluation of flocculant dosing methods to minimise cell lysis is required to optimise the process.

Membrane filtration is also possible, both to remove cells and also to remove toxins. Testing of appropriate membrane systems is underway by manufacturers.

When water treatment plants are distant from the supply reservoir, cell lysis may occur in the pipeline, especially if pressure reduction is applied. In conventional plants pre-chlorination will lyse cyanobacterial cells. The outcome of both is that soluble toxins are released into the water, and require removal during treatment. The early research on toxin removal focussed on the use of activated carbon, since it was already in use for taste and odour removal, and removal of undesirable organic compounds from drinking water. Laboratory work, and pilot plant tests, showed that both filtration through granular activated carbon, and injection of powdered activated carbon are capable of removal of cyanobacterial toxins [175, 183, 184]. Considerable differences in adsorbtion capability were seen between different sources of activated carbon, in adsorbtion of microcystins and nodularin [185]. Activated carbon thus can be effective but is costly, due to the cost of supply of powdered carbon, or granular carbon, plant modifications, monitoring effectiveness, and replacement of exhausted granular carbon.

Oxidation of toxins by excess chlorine, potassium permanganate and ozone are alternatives being tested [186]. Ozone appears the most promising oxidant for drinking water supplies, and is under extensive evaluation [187–190]. Ultraviolet light will at low intensities isomerise ADDA, and at high intensities cleave the microcystin peptide ring and render the compound non-toxic [17]. The level of UV irradiation required may, however, be too high to be used in public water supply [17, 188].

Because of the relatively recent discovery of paralytic shellfish poisons in *Anabaena* [80], little work has been done on water treatment methods for removal of these toxins. Activated carbon has, however, been shown to be effective with *Anabaena* neurotoxin from an Australian source, which has since been shown to be paralytic shellfish poison [184].

6
Reservoir Treatment

Many water supply authorities, and owners of private supplies, regularly use copper sulphate dosing to suppress or kill cyanobacterial blooms. The standard treatment is to apply a copper concentration of 1 ppm in the top metre of water in a reservoir. This is applied from a boat, helicopter or aircraft depending on the area to be treated. Such treatment causes lysis of cyanobacterial cells, and release of toxins into the free water mass [191]. Both of the public outbreaks of poisoning due to cyanobacterial toxins in Australia have followed reservoir treatment with copper, causing lysis of a cyanobacterial bloom [10, 24].

Use of other algicides is similarly complicated by the release of toxin [182]. It is apparent that cyanobacterial control by algicide treatment can only be safely used at very low cell densities, or in conjunction with diversion of the supply to another water source.

Toxins freed into lakes on cell lysis degrade through bacterial decomposition [192]. Microcystins are persistent molecules in the environment, retaining toxicity in dried crusts of cyanobacteria for months [193]. However, natural degradation in water is observed [194].

Use of copper in aquatic environments is controlled in many countries. In Australia copper treatment of artificial water reservoirs is allowed, whereas use of copper in natural lakes and rivers is prohibited. A more environmentally acceptable method of cyanobacterial control in water storages is through the use of aeration and destratification [195].

7
Risk Estimations

The only class of cyanobacterial toxins for which there is sufficient data for risk estimation in water supplies at the present time is the cyclic peptide hepatotoxins [18, 146]. These have been discussed earlier, (Sect. 3.2.1.4) and so far the major issue identified affecting uncertainty factors to be applied to this experimental toxicity data is that of tumour promotion [132, 138]. However, the use of sub-chronic animal data for human risk estimation involves a sequence of assumptions, which lead to uncertainty factors which may be widely inaccurate [18, 146]. Cyanobacteria also contain a very wide variety of other biologically active compounds, including phorbol-ester type tumour promoters, cytotoxins, and immunosuppressive compounds. It can be expected that the current search for potential pharmaceutical products from cyanobacteria will uncover a range of compounds capable of harm to the human population [196].

To resolve the question of risk to populations from cyanobacteria, large scale epidemiological investigation is needed to obtain correlations between human cyanobacterial exposure and persistent health impacts. Evaluation of exposed populations for birth defects, tumour incidence, mortality and morbidity is required to clarify the overall health significance of toxic cyanobacteria in drinking and recreational waters.

End-note

Recent discussion by two Working Groups of the World Health Organisation, based on the data given in Section 3.2.1.5 on page 70, have resulted in the recommendation of a Guideline Value for microcystin-LR in drinking water of 1 μg l^{-1} [197].

Acknowledgements. I would like to thank Mrs Shirley Pearce for excellent secretarial help and Mr Andrew Humpage for bibliographic assistance.

References

1. Skulberg OM, Carmichael WW, Codd GA, Skulberg R (1993) In: Falconer IR (ed) Algal toxins in seafood and drinking water. Academic Press, London, p 145
2. Bowling L (1994) Aust J Marine Freshwater Res 45:737
3. Baker PD, Humpage AR (1994) Aust J Marine Freshwater Res 45:773
4. Lawton LA, Codd GA (1991) J Inst Water Env Mgt 5:460
5. Francis G (1878) Nature (London) 18:11
6. Carmichael WW, Falconer IR (1993) In: Falconer IR (ed) Algal toxins in seafood and drinking water. Academic Press, London, p 187
7. Turner PC, Gammie AJ, Hollinrake LT, Codd GA (1990) Br Med J 300:1440
8. Lippy EC, Erb J (1976) J Am Water Works Assoc 68:606
9. Bourke ACT, Hawes RB, Nielson A, Stallman ND (1983) Toxicon Suppl 3:45
10. Hawkins PR, Runnegar MTC, Jackson ARB, Falconer IR (1985) Appl Env Microbiol 50:1292
11. Vollenweider RA (1968) Scientific fundamentals of the eutrophication of lakes and flowing waters with particular reference to nitrogen and phosphorus as factors in eutrophication, OECD, Paris
12. Reynolds CS, Walsby AE (1975) Biol Rev 50:437
13. Sivonen K (1990) Appl Environ Microbiol 56:2658
14. Falconer IR, Runnegar MTC, Huyn VL, Bradshaw WP (1989) J Am Water Works Assoc 81:102
15. Keijola AM, Himberg K, Esala K, Sivonen K, Hiisvirta L (1988) Toxic Assess 3:643
16. Himberg K, Keijola AM, Hiisvirta L, Pyysalo H, Sivonen K (1989) Water Res 23:979
17. Tsuji K, Watanuki T, Kondo F, Watanabe MF, Suzuki S, Nakazawa H, Suzuki M, Uchida H, Harada K-I (1995) Toxicon 33:1619
18. Falconer IR, Burch MD, Steffensen DA, Choice M, Coverdale OR (1994) Env Toxicol Water Qual 9:131
19. Falconer IR, Humpage AR (1996) Phycologia, 35:(6 Suppl) 6
20. Chu FS, Huang X, Wei RD, Carmichael WW (1989) Appl Environ Microbiol 55:1928
21. Chu FS, Huang X, Wei RD (1990) J Assoc Off Analyt Chem 73:451
22. Holmes CFB (1991) Toxicon 29:469
23. Lambert TW, Boland, MP, Holmes CFB, Hrudey SE (1994) Environ Sci Technol 28:753
24. Falconer, IR, Beresford AM, Runnegar MTC (1983) Med J Aust 1:511
25. El Saadi O, Esterman AJ, Cameron S, Roder D (1995) Med J Aust 162:122

26. Yu S-Z (1994) In: Steffensen DA, Nicholson BC (eds) Toxic cyanobacteria, current status of research and management. Australian Centre for Water Quality Research, Salisbury, Australia, p 75
27. Yu S-Z (1995) J Gastroenterol Hepatol 10:674
28. Rouhiainen L, Sivonen K, Buikkema WJ, Haselkorn R (1995) J Bacteriol 177:6021
29. Neilan BA, Hawkins PR, Cox PT, Goodman A (1994) Aust J Marine Freshwater Res 45:869
30. Scott WE, (1991) Wat Sci Tech 23:175
31. Watanabe M (1987) Bull Natu Sci Mus Tokyo B13:81
32. Repavich WM, Sonzogni WC, Standridge JH, Wedepohl RE, Meisner L (1990) Wat Res 24:225
33. Sivonen K, Niemela SI, Niemi RM, Lepisto L, Luoma TH, Räsänen LA (1990) Hydrobiologia 190:267
34. Pearson MJ (1990) Toxic blue-green algae. Report of the National Rivers Authority. Water Quality Series No 2. National Rivers Authority, London, p 27
35. Baker PD, Humpage AR (1994) Aust J Marine Freshwater Res 45:773
36. Kotek BG, Lam AK-Y, Prepas EE, Kenefide SL, Hrudey SE (1995) J Phycol 31:248
37. Rouhiainen L, Buikkema W, Paulin L, Sivonen K, Haselkorn R (1996) Hydrobiologia. In Press
38. Meissner K, Dittmann E, Börner T (1996) FEMS Microbiol Lett 135:295
39. Codd GA, Poon GK (1988) In: Gallon JG (ed) Biochemistry of the algae and cyanobacteria. Oxford Scientific Publ, Clarendon, Oxford, p 283
40. Watanabe MF, Oishi S (1985) Appl Environ Microbiol 49:1342
41. Sivonen K (1990) Appl Environ Microbiol 56:2658
42. Van der Westhuizen AJ, Eloff JN (1985) Planta 163:55
43. Ganf GG, Oliver RL (1987) J Ecol 70:829
44. Reynolds CS (1987) Adv Biol Res 13:67
45. Falconer IR, Jackson ARB, Langley J, Runnegar MTC (1981) Aust J Biol Sci 34:179
46. Carmichael WW, Mahmood MA, Hyde EG (1990) In: Hall S, Strichartz (eds) Marine toxins:origin, structure and molecular pharmacology. American Chemical Society, Washington DC, p 87
47. Davidson FF (1959) Am J Water Works Assoc 51:1277
48. Sivonen K, Himberg K, Luukkainen R, Niemela SI, Poon GK, Codd GA (1989) Toxic Assess 4:339
49. Skulberg OM, Carmichael WW, Andersen RA, Matsunaga S, Moore RE, Skulberg R (1992) Environ Toxicol Chem 11:321
50. Edwards C, Beattie KA, Scrimgeour CM, Codd GA (1992) Toxicon 30:1165
51. Hawser SP, Codd GA, Capone DG, Carpenter EJC (1991) Toxicon 29:277
52. Falconer IR (1997) Personal communication
53. Huber CS (1972) Acta Crystallograph 328:2577
54. Devlin JP, Edwards OE, Gorham PR, Hunter NR, Pike RK, Stavric B (1977) Can J Chem 55:1367
55. Harada K-I, Kimura Y, Ogawa K, Suzuki M, Dahlem AM, Beasley VR, Carmichael WW (1989) Toxicon 27:1289
56. Himberg K (1989) J Chromatogr 481:358
57. Stevens DK, Krieger RI (1988) J Analyt Toxicol 12:126
58. Ojanperä I, Vuori E, Himberg K, Waris M, Niinivaara K (1991) Analyst 116:265
59. Gallon JR, Chit KN, Brown EG (1990) Phytochem 29:1107
60. Carmichael WW, Biggs DF, Gorham PR (1975) Mol Pharmacol 18:384
61. Aronstam RS, Witkop B (1981) Proc Natl Acad Sci USA 78:4639
62. Gorham PR, Carmichael WW (1988) In: Lembi CA, Waaland JR (eds) Algae and human affairs. Cambridge University Press, Cambridge, p 403
63. Carmichael WW (1992) J Appl Bacteriol 72:445
64. Jones GJ, Korth W (1995) Wat Sci Tech 31:145
65. Carmichael WW, Gorham PR (1978) Mitteil Int Verein Limnol 21:285

66. Matsunaga S, Moore RE, Niemczara WP, Carmichael WW (1989) J Amer Chem Soc 111:8021
67. Falconer IR (1993) In: Falconer IR (ed) Algal toxins in seafood and drinking water. Academic Press, London, p 165 and p 177
68. Wills JH (1982) Lab Management 20:53
69. Moore BS, Ohtani I, de Koning CB, Moore REC (1992) Tetrahedron Lett 33:6595
70. Mahmood NA, Carmichael WW (1986) Toxicon 24:425
71. Cook WO, Beasley VR, Lovell RA, Dahlem AM, Hooser SB, Mahmood NA, Carmichael WW (1989) Environ Toxicol Chem 8:915
72. Kao CY (1993) In: Falconer IR (ed) Algal toxins in seafood and drinking water. Academic Press, London, p 75
73. Sawyer PJ, Gentile JH, Sasner JJ (1968) Can J Microbiol 14:1199
74. Jackim E, Gentle J (1968) Science (Washington DC) 162:915
75. Alam M, Shimiza Y, Ikawa M, Sasner JJ (1978) J Environ Sci Health A13:493
76. Sasner JJ, Ikawa M, Foxall TL (1984) In: Ragelis EP (ed) Seafood toxins. American Chemical Society, Washington DC, p 391
77. May V, McBarron EJ (1973) J Aust Inst Agric Sci 39:264
78. Runnegar MTC, Jackson ARB, Falconer IR (1988) Toxicon 26:599
79. Bowling L (1992) The cyanobacterial (blue-green algal) bloom in the Darling/Barwon River system, November-December 1991. NSW Department of Water Resources, Technical Services Division, Report No 92.074
80. Humpage AR, Rositano J, Bretag AH, Brown R, Baker PD, Nicholson BC, Steffensen DA (1994) Aust J Marine Freshwater Res 45:761
81. Sullivan JJ (1993) In: Falconer IR (ed) Algal toxins in seafood and drinking water. Academic Press, London, p 29
82. Baden DG, Trainer VL (1993) In: Falconer IR (ed) Algal toxins in seafood and drinking water. Academic Press, London, p 49
83. Ohsima Y, Sugino K, Yasumoto T (1989) In: Phycotoxins '88. Elsevier, Amsterdam, p 319
84. Negri AP, Jones GJ (1995) Toxicon 33:667
85. Steyn DG (1945) S African J Sci 41:343
86. Senior VE (1960) Can J Comp Med 24:36
87. McBarron EJ, May V (1966) Aust Vet J 42:449
88. Carmichael WW, Beasley V, Bunner DL, Eloff JN, Falconer IR, Gorham PR, Harada K-I, Yu M-J, Krishnamurthy T, Moore RE, Rinehart K, Runnegar M, Skulberg OM, Watanabe M (1988) Toxicon 26:971
89. Rinehart KL, Harada K-I, Namikoshi M, Chen C, Harvis CA, Munro MHG, Blunt JW, Mulligan PE, Beasley VR, Dahlem AM, Carmichael WW (1988) J Am Chem Soc 110:8557
90. Oxenford J (1996) Am Water Works Assoc Research Foundation. Personal communication
91. Pilotto, L (1997) National Centre for Epidemiology and Public Health, Canberra. Personal communication
92. Watanabe M (1996) In: Watanabe MF, Harada K-I, Carmichael WW, Fujiki (eds) Toxic microcystis. CRC, Tokyo, p 13
93. Honkanen RE, Caplan FR, Baker KK, Baldwin CL, Bobzin SC, Bolis CM, Cabrera GM, Johnson LA, Jung JH, Larsen LK, Levine IA, Moore RE, Nelson CS, Patterson GML, Tschappat KD, Tuang GD, Boynton AL, Arment AR, An J, Carmichael WW, Rodland KD, Magun BE, Lewin RA (1995) J Phycol 31:478
94. Botes DP, Tuiman AA, Wessels PL, Viljoen CC, Kruger H, Williams DH, Santikarn S, Smith RJ, Hammond SJ (1984) J Chem Soc Perkin Trans 1:2311
95. Elleman TC, Falconer IR, Jackson ARB, Runnegar MTC (1978) Aust J Biol Sci 31:209
96. Botes DP, Wessels PL, Kruger H, Runnegar MTC, Santikarn S, Smith RJ, Barna JCJ, Williams DH (1985) J Chem Soc Perkin Trans 1:2747
97. Dahlem AM (1989) Structure/toxicity relationships and fate of low molecular weight peptide toxins from cyanobacteria. PhD thesis, Univ of Illinois, USA

98. Harada K-I, Ogawa K, Matsaura K, Murata H, Suzuki M, Watanabe MF, Ikezono Y, Nakayama N (1990) Chem Res Toxicol 3:473
99. Harada K-I (1996) In: Watanabe MF, Harada K-I, Carmichael WW, Fujiki (eds) Toxic microcystis. CRC, Tokyo, p 103
100. Namikoshi M, Rinehart KL, Dahlem AM, Beasley VR, Carmichael WW (1989) Tetrahedron Lett 30:4349
101. Moore RE, Chen JL, Moore SB, Patterson GML (1991) J Am Chem Soc 113:5083
102. Rinehart KL, Namikoshi M, Choi BM (1994) J Appl Phycol 6:159
103. Falconer, IR (1993) In: Falconer IR (ed) Algal toxins in seafood and drinking water. Academic Press, London, p 165
104. Kusumi T (1996) In: Watanabe MF, Harada K-I, Carmichael WW, Fujiki (eds) Toxic microcystis. CRC, Tokyo, p 103
105. Chu FS, Huang X, Wei RD, Carmichael WW (1989) Appl Environ Microbiol 55:1928
106. Chu FS, Huang X, Wei RD (1990) J Assoc Off Analyt Chem 73:451
107. An J-S, Carmichael WW (1994) Toxicon 32:1495
108. McDermott CM, Feola R, Plude J (1995) Toxicon 33:1433
109. MacKintosh C, Beattie KA, Klumpp C, Cohen P, Codd GA (1990) FEBS Lett 264:187
110. Honkanen RE, Zwiller J, Moore RE, Daily SL, Khatra BS, Dukelow M, Boynton AL (1990) J Biol Chem 265:19, 401
111. Yoshizawa S, Matsushima R, Watanabe MF, Harada K, Ichihara A, Carmichael WW, Fujiki H (1990) J Cancer Res Clin Oncol 116:609
112. Kaya K (1996) In: Watanabe MF, Harada K-I, Carmichael WW, Fujiki H (eds) Toxic microcystis. CRC Press Tokyo, p 175
113. Ash C, MacKintosh C, MacKintosh R, Fricker CR (1995) Wat Sci Tech 31:47
114. Ash C, MacKintosh C, MacKintosh R, Fricker CR (1995) Wat Sci Tech 31:51
115. Lambert TW, Boland MP, Holmes CFB, Hrudey SE (1994) Environ Sci Technol 28:753
116. Jackson ARB, McInnes A, Falconer IR, Runnegar MTC (1984) Vet Pathol 21:102
117. Brooks WP, Codd GA (1987) Pharmacol Toxicol 60:187
118. Falconer IR, Buckley T, Runnegar MTC (1986) Aust J Biol Sci 39:17
119. Eriksson JE, Hägerstrand H, Isomaa B (1987) Biochim Biophys Acta 930:304
120. Dahlem AM, Hassan AS, Swanson SP, Carmichael WW, Beasley VR (1988) Pharmacol Toxicol 63:1
121. Falconer IR, Dornbusch M, Moran G, Yeung SK (1992) Toxicon 30:790
122. Runnegar MT, Falconer IR (1982) S Africa J Sci 78:363
123. Eriksson JE, Gröberg L, Nygärd S, Slate YP, Meriluoto JAO (1990) Biochim Biophys Acta 1025:60
124. Runnegar MTC, Gerdes RG, Falconer IR (1991) Toxicon 29:45
125. Theiss WC, Carmichael WW, Wyman J, Bruner R (1988) Toxicon 26:603
126. Runnegar MT, Falconer IR, Silver J (1981) Naunyn-Schmiedeberg's Arch Pharmacol 317:268
127. Eriksson JE, Paatero GIL, Meriluoto JAO, Codd GA, Kass GEN, Micotera P, Orrenius S (1989) Exp Cell Res 185:86
128. Falconer IR, Yeung DSK (1992) Chem Biol Interact 81:181
129. Wickstrom ML, Khan SA, Haschek WM, Wyman JF, Eriksson JE, Schaeffer DJ, Beasley VR (1995) Toxicol Pathol 23:326
130. Falconer IR (1993) In: Falconer IR (ed) Algal toxins in seafood and drinking water. Academic Press, London, p 177
131. Chou Y-H, Rosevear E, Goldman RD (1989) Proc Natl Acad Sci 86:1885
132. Fujiki H, Sueoka E, Suganuma M (1996) In:Watanabe MF, Harada K-I, Carmichael WW, Fujiki H (eds) Toxic Microcystis. CRC Press, Tokyo, p 203
133. Carbis CR, Simons JA, Mitchell GF, Anderson JW, McCauley I (1994) Res Vet Sci 57; 310
134. Slatkin DN, Stoner, RD, Adams WH, Kycia JH, Siegelman WH (1983) Science 220:1383
135. Falconer IR, Smith JV, Jackson ARB, Jones A, Runnegar MTC (1988) J Toxicol Environ Health 24:291

136. Carmichael WW, An J-S, Azevedo SMFO, Lau S, Rinehart KL, Jochimsen EM, Holmes CEM, Jarvis WR (1996) Analysis for microcystins involved in an outbreak of liver failure and death of humans at a hemodialysis center in Caruaru, Pernambuco Brazil. IV Symposium of the Brazilian Society of Toxinology – October 6–12, Recife, Pernambuco, Brazil

137. Falconer IR, Buckley TH (1989) Med J Aust 150:351

138. Falconer IR (1991) Environ Toxicol Water Qual 6:177

139. Nishiwaki-Matsushima R, Ohta T, Nishiwaki S, Suganuma M, Kohyama K, Ishikawa T, Carmichael WW, Fujiki H (1992) J Cancer Res Clin Oncol 118:420

140. Yuasa H, Yoshida K, Iwata H, Nakaishi H, Suganuma M, Tatematsu MC (1994) J Cancer Res Clin Oncol 120:208

141. Ohta T, Nishiwaki R, Yatsunami J, Komori A, Suganuma M, Fujiki H (1992) Carcinogenesis 13:2443

142. Ashendel CL, Baudoin PA, Minor PL (1988) In: Langenback R, Elmore E, Barrett JC (eds) Progress in Cancer Research and Therapy 34: Tumor promoters: biological approaches for mechanistic studies and assay systems. Raven Press, New York, p 331

143. Nakano M, Nakano Y, Sato-Taki T, Mari N, Kojima M, Ohtake A, Shinai M (1989) Microbiol Immunol 33:787

144. Nakano Y, Shinai M, Mari N, Nakano M (1991) Appl Environ Microbiol 57:327

145. Naseem SM, Mereish KA, Solow R, Hines HB (1991) Toxic In Vitro 5:341

146. Kuiper-Goodman T, Gupta S, Combley H, Thomas BH (1994) In: Steffensen DA, Nicholson BC (eds) Toxic Cyanobacteria:current status of research and management. Australian Centre for Water Quality Research, Salisbury, Australia, p 67

147. Codd GA, Steffensen DA, Burch MD, Baker PD (1994) Aust J Marine Freshwater Res 45:731

148. Jones GJ, Blackburn SI, Parker NS (1994) Aust J Marine Freshwater Res 45:787

149. Edler L, Ferno S, Lind MG, Lundberg R, Nilsson PO (1985) Ophelia 24:103

150. Gussmann H, Molzahn MJ, Bicks B (1985) Monatshefte für Veterinärmed 40:76

151. Sivonen K, Kononen K, Esala A-L, Niemelä S1 (1987) Hydrobiologia 183:3

152. Kononen K (1992) Finnish Marine Res 261:3

153. Kononen K, Sivonen K, Lehtimäki J (1993) In: Smayda TJ, Shimiza Y (eds) Toxic phytoplankton blooms in the sea. Elsevier, Amsterdam

154. Eriksson JE, Meriluoto JAO, Jujari HP, Osterlund K, Fagerlund K, Hällbom L (1988) Toxicon 26:161

155. Runnegar MTC, Jackson ARB, Falconer IR (1988) Toxicon 26:143

156. Ohta T, Sueoka E, Iida N, Komori A, Suganuma M, Nishiwaki R, Tatematsu M, Kim SJ, Carmichael WW, Fujiki H (1994) Cancer Res 54:6402

157. McComb AJ, Humphries R (1992) Estuaries 15:529

158. Kinhill Engineers Pty (1988) Peel Inlet and Harvey Estuary Management Strategy: Environmental Review and Management Programme, Stage 2. Departments of Agriculture; Marine and Harbours, Western Australia

159. Falconer IR, Choice A, Hosja WJ (1993) Env Toxicol Water Qual 7:119

160. de Silva ED, Williams DE, Andersen RJ, Klix H, Holmes CFB, Allen TM (1992) Tetrahedron Lett 33:1561

161. Ohtani I, Moore RE, Runnegar MTC (1992) J Am Chem Soc 114:7941

162. Harada K-I, Ohtani I, Iwamoto K, Suzuki M, Watanabe ME, Watanabe M, Terao K (1994) Toxicon 32:73

163. Terao K, Ohmori S, Igareshi K, Ohtani I, Watanabe, MF, Harada KI, Ito E, Watanabe M (1994) Toxicon 32:835

164. Runnegar M, Kong S, Zhong Y, Ge J, Lu S (1994) Biochem Biophys Res Commun 201:235

165. Runnegar MTC, Kong S-M, Zhang Y-Z, Lu SC (1995) Biochemical Pharmacol 49:319

166. Hawkins PR, Chandrasena MR, Jones GJ, Humpage AR, Falconer IR (1997) Isolation and toxicity of *Cylindrospermopsis raciborskii* from an ornamental lake. Toxicon 35:341–346

167. Byth S (1980) Med J Aust 2:40
168. Bourke ATC, Hawes RB, Neilson A, Stallman ND (1983) Toxicon Supp 13:45
169. Hayman J (1992) Med J Aust 157:794
170. Prociv P (1993) Med J Aust 158:434
171. Falconer IR (1993) Med J Aust 158:434
172. Hiese HA (1949) J Allergy 20:383
173. Hiese HA (1951) Ann Allergy 9:100
174. Cohen SG, Reif OB (1953) J Allergy 24:452
175. Mittal A, Argarwal MK, Schivpuri DN (1979) Ann Allergy 42:253
176. Grauer FH (1961) Arch Dermatol 84:720
177. Izumi AK, Moore RE (1987) Clinics Dermatol 5:92
178. Moore RE, Patterson, GML, Entzeroth M, Morimoto H, Suganuma M, Hakii, H, Fujiki H,
     Sugimura T (1986) Carcinogenesis 7:641
179. Gorham PR, Carmichael WW (1988) In: Lembi CA, Waaland JR (eds) Algae and human
     affairs. Cambridge Univ Press, New York, p 403
180. Soong FS, Maynard E, Kirke E, Luke C (1992) Med J Aust 156:67
181. Falconer IR (1997) Personal communication
182. Lam AK-Y, Prepas EE, Spink D, Hrudey SE (1995) Wat Res 29:1845
183. Hoffmann JRH (1976) Water SA 2:58
184. Falconer IR, Runnegar MTC, Buckley T, Huyn VL, Bradshaw P (1978) J Am Water Works
     Assoc 81:102
185. Donati C, Drikas M, Hayes R, Newcombe G (1994) Wat Res 28:1735
186. Rositano J, Nicholson BC (1994) Water treatment techniques for the removal of cyano-
     bacterial peptide toxins from water. Australian Centre for Water Quality and Treatment
     Report 2/94, Salisbury, Australia
187. Himberg K, Keijola AM, Hiisvirta L, Pyysalo K, Sivonen K (1989) Wat Res 23:979
188. Drikas M (1994) In: Steffensen DA, Nicholson BC (eds) Toxic cyanobacteria: current
     status of research and management. Australian Centre for Water Quality Research,
     Salisbury, Australia, p 93
189. Keijola AM, Himberg K, Esala AL, Sivonen K, Hiisvirta L (1988) Toxicity Assess 3:643
190. James H, Fawell JK (1991) In:Detection and removal of cyanobacterial toxins from
     freshwaters. Foundation for Water Research, Medenham UK Report FR 0211
191. Kenetick SL, Hrudey SE, Peterson HG, Prepas EE (1993) Wat Sci Technol 27:433
192. Rapala J, Lahti K, Sivonen K, Niemela SI (1994) Lett Appl Microbiol 19:423
193. Jones GJ, Falconer IR, Wilkins RM (1995) Environ Toxicol Wat Quality 10:19
194. Jones GJ, Orr PT (1994) Wat Res 28:871
195. Yoo RS, Carmichael WW, Hoehn RC, Hrudey SE (1995) Cyanobacterial (Blue-Green
     Algal) Toxins: A Resource Guide. American Water Works Association Research
     Foundation, Denver, Colorado, p 122
196. Falch BS, König GM, Wright AD, Sticher O, Angerhofer CK, Pezzuto JM, Bachman H
     (1995) Planta Med 61:321
197. WHO Working Group Meeting on Chemical Substances in Drinking Water (1997)
     Geneva, Report Section 5.2

Chemistry of Aqueous Ozone and Transformation of Pollutants by Ozonation and Advanced Oxidation Processes

Jürg Hoigné

Swiss Federal Institute for Environmental Science and Technology (EAWAG),
CH-8600 Dübendorf, Switzerland
E-mail: hoigne@eawag.ch

Ozonation is widely and successfully applied for many types of oxidative water treatments. Its chemical effects can be described by considering the sequences of highly selective direct reactions of molecular ozone and the reactions of the more reactive but less selective OH radicals which are always produced from decomposed ozone in aqueous systems. These radicals also control the ozone based AOPs (Advanced Oxidation Processes). In some cases even formation and reactions of additional secondary oxidants, such as carbonate radicals, hypobromite, and hydrogen peroxide have to be accounted for.

There is a lot of practical experience and basic data on reaction rates as well as on potential product formations available to predict chemical transformations that occur during ozonation processes. However, in order to apply this background, one must have information on the concentration and lifetime of the dissolved ozone as well as of some relevant secondary oxidants. Such parameters strongly depend on the dosage of ozone and on the chemical composition of the water. In addition, the hydraulics of the reactor used for the ozonation are relevant when we want to relate disinfection goals or transformation of micropollutants or by-product formation with ozone dosages. Qualification of the treated water and of the ozonation by-product in respect to drinkingwater quality or environmenatal aspects requires that subsequent downstream processes, such as microbiological post-treament, post-chlorination, and reactions in the distribution system are also considered.

Keywords: drinking water, hydroxyl radical, ozone, reaction kinetics, water treatment, advanced oxidation process (AOP).

Contents

The Handbook of Environmental Chemistry Vol. 5 Part C
Quality and Treatment of Drinking Water II (ed. by J. Hrubec)
© Springer-Verlag Berlin Heidelberg 1998

List of Symbols and Abbreviations

AC   activated carbon
AOC   assimilable organic carbon
AOP   advanced oxidation process
c · t-value   oxidant exposure value, i.e. conc. of oxidant (c) integrated for
                reaction time t
DBP   disinfection by-products
DNOM   dissolved natural organic material
DOC   dissolved organic carbon
DOM   dissolved organic material
GAC   granular activated carbon

HO_2/O_2^-	hydroperoxyl radical and superoxide anion ("dioxidanyl")
$k_{OH,P}$	second-order reaction-rate constant for $OH^\cdot$ reacting with a target compound P
$k_{OH,S}$	second-order reaction-rate constant for $OH^\cdot$ reacting with a scavenger compound S
IOA	International Ozone Association
M	molar (mol/dm^3), or: specified molecular species
P	specified molecular trace pollutant or probe compound
(ΔO_3)	amount of ozone per dm^3 that has decomposed in a water
$(\Delta O_3)_{37\%}$,	(ΔO_3) required to decrease the conc. of a target molecule P by a factor of e (i.e. to 37%). Also called Ω-value.
$OH^\cdot$ (or $HO^\cdot$)	hydroxyl radical
S	scavenger for $OH^\cdot$ or for carbonate radicals
TOC	total organic carbon
TOX	total organic halogen compounds
[X]t	conc. of compound X at time t (M)
[X](t)	conc. of compound X as a function of time t (M)
η	stoichiometric yield factor (mole/mole)
τ_x	lifetime of a species X. Within this time the conc. of X declines by a factor e, to 37%.

1
Introduction

1.1
Aim of this Survey

For more than three decades, ozone has been widely applied in many regions of the world for the treatment of waters including drinking water, industrial waste waters, cooling water, swimming pools and marine aquaria. The process goals are thereby to disinfect, to improve color, taste and odor, to oxidize reduced manganese and iron species, to degrade and oxidize organic compounds, to enhance the microbiological degradability of organic compounds such as of natural organic material (NOM), or even to improve succeeding flocculation and filtration processes. However, some oxidation by-products are of concern not only in respect to their direct human toxicity and their potential to produce suspicious secondary products, but also in respect to detrimental effects that occur during subsequent treatment steps and water distribution.

A rapidly increasing number of useful research papers and reports on the application of ozone for water treatment have been released. At present more than hundred papers a year are submitted that reflect the many projects underway. However, up-dated reviews have been issued regularly to help sift through all this information [1–4]. Also many different aspects of applied ozonation have been well evaluated to assess further research needs in water treatment [5, 6]. Further actualities that contribute to the evolution of the field are reported in proceedings of conferences such as the World Conferences of the International Ozone Association (for an example cf. [7]), or in specialized journals such as the *Journal of the International Ozone Association* (IOA), *Ozone, Science and Engineering* [8], that is complemented by *Ozone News* [9]. Information can also be found in all journals that focus on water treatment and environmental technology. Further insight can be also be achieved from basic ozone chemistry as of interest in organic chemistry (e.g. [10, 11], and from publications on reactions of ozone and AOPs that occur in the natural aqueous environment, such as in clouds (for some references cf. [12]). In addition, up-dated compilations of rate constants for reaction in aqueous solutions now also list data for 250 reactions of aqueous ozone, 1500 for OH radicals, and 200 for carbonate radicals. These also include data for many further secondary oxidants [13, 14].

Chemists, chemical engineers and hygienists, however, require a basic framework that allows them to critically select and allocate the compiled information. For this, basic chemistry based on reaction kinetic concepts has been shown to be a rather useful help. When we can delineate the complex network of pathways of ozonation processes and isolate individual basic reactions of unique rate constants and specified product formations, we can also apply the wealth of basic chemical information that has already been compiled by scientists working in complementary fields. Comprehensive kinetic chemical models can then be composed that are suitable for the application of kinetic computer programs and that can be tested and calibrated by better focused experiments.

Section 1 (Introduction) includes a summary of treatment goals and some data on ozone (O_3) and treatment processes. Section 2 outlines the direct molecular reactions of O_3 with dissolved compounds and compares reaction-rate constants for compounds of particular interest. Section 3 then shows how O_3 is transformed into secondary oxidants, such as hydroxyl radicals (OH$^\cdot$), hydroperoxyl radicals (HO$_2^\cdot$), and further species (O$_3^-$, HO$_3^\cdot$, etc.). Also included is a discussion of the kinetics of the involved radical-type chain reactions which often control the lifetime of O_3 in water. Section 4 compares different methods to accelerate the formation of OH$^\cdot$ that initiates ozone based Advanced Oxidation Processes (AOPs). In Sect. 5 the role of OH$^\cdot$ for oxidizing further pollutants is quantified. A formulation of the competition kinetics involved enables different types of waters to be qualified with regard to their ability to promote, accelerate, or inhibit such ozone based radical-type oxidation processes. Section 6 starts by showing that the efficiency of an ozonation process and its by-product formation also depends on the hydraulics of the reactors. Then the formation of typical by-products is reviewed. We can then learn that, in exceptional cases, even complex chains of product formations can be predicted by computer simulations, provided that comprehensive sets of data are available for the kinetics of all relevant reactions as well as on the composition of the water, and on the hydraulic of the reactor. The most advanced and successful example to demonstrate this is the formation of bromate in bromide containing water. We can then validate the present information on the formation of by-products that are produced during drinking water ozonation and discuss their health effects. Effects of conventional process combinations on product formation and water quality improvements are reviewed in Sect. 7. Section 8 summarizes the conclusions.

1.2
Treatment Goals and Applications

1.2.1
Areas of Application of Ozonation Processes

– drinking water
– cooling water
– swimming pools
– bottled water
– industrial waters containing phenols, cyanides, …
– leakage from landfills
– exhaust air washing-water
– in-plant water lines for ultra pure waters
– marine aquaria waters (based on bromide free salts)

1.2.2
The Main Effects of Ozonation Treatment

– disinfection
– oxidations of
  • Mn(II), Fe (II), …(e.g. present in reduced ground waters)
  • phenol, chlorophenol, amines, olefinic compounds
  • cyanide
  • bromide ion
  • colors and taste forming compounds
– improvement of the microbiological degradation of dissolved organic materials in succeeding microbiological processes
– possible improvement of succeeding sedimentation, flocculation-filtration and flotation processes.

1.2.3
Drawbacks of Concern

– aggressive odor of ozone residual
– short lifetime of ozone in water (in oxidized waters: a few minutes, up to an hour)
– Formation of ozonation by-products:
  • microbiologically degradable organic material from the oxidation of dissolved (natural) organic material (DNOM) that lead to enhanced microbiological fouling in the distribution system
  • bromate (when bromide is present)
  • brominated organic compounds (when bromide and DOM is present)
  • permanganate (if Mn^{2+} is present)
  • aldehydes, organic acids and further carbonyl compounds derived from oxidized DNOM.

1.2.4
Typical Dosages of Applied Ozone

- 1 to 3 mg/l ozone dosage for conventional drinking water production from "good raw waters"; often also a dosage of 1–2 mg/l of ozone per mg/l of DOC has been recommended;
- higher dosage for waste water treatment (consider stoichiometry);
- higher dosages when ozone is transformed into non-selctive OH' (enhanced in AOPs);
- lower dosage (0.4 to 0.8 mg/mg of DOC) when O_3 is just applied as a coagulant or as a disinfectant or when by-product formation (such as formation of bromate) are of concern.

1.2.5
Ozone Application

- gas-exchange is used from air or oxygen that typically contains 2 to 6 (or even more) v/v % of O_3 (produced on site) (compare entries in Fig. 1)

1.3
Data for Ozone

- chemical structure: O_3 is a singlet diradical in its groundstate. Its structure has been described as a resonance hybrid of four canonical forms of dipolar character that mark a continuous scale. Its electrons remain paired (for further details and references cf. [10]).

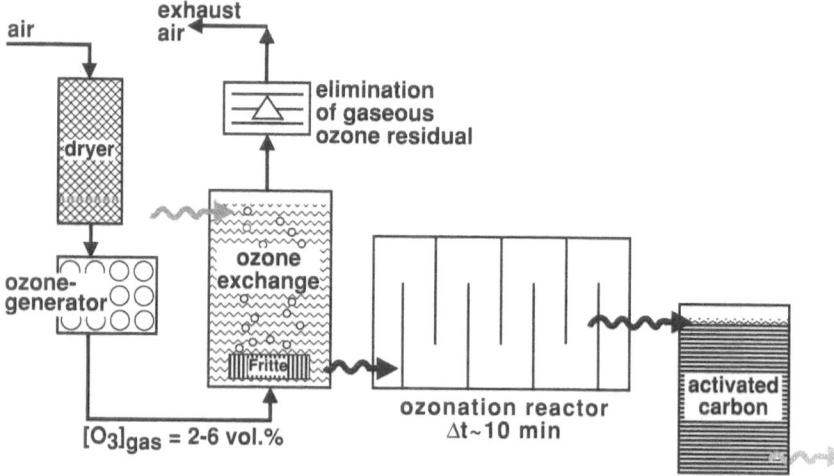

Fig. 1. Example for an ozonation process-line

- molecular mass: 48.0 g
- melting point: −193 °C
- UV absorption band in water:

ε_{max} at 258 nm: 3150 $M^{-1} cm^{-1}$ [15] (this broad absorption band extends up into the region of 300 nm)

- visible absorption band:

ε_{max} at 600 nm: about 1.2 $M^{-1} cm^{-1}$ (calc. from ε_{gas}); concentrated solutions of O_3 appear blue

- Henry constants:

K_H (0 °C): 35 atm/(mol/l);
K_H (20 °C): 100 atm/(mol/l);

- instant odor threshold conc. in air: about 40 μg m^{-3}, but adaptation takes place within minutes [16];
- Maximal allowable conc. of ozone in air:

< 200 μg m^{-3} air (about 0.1 ppmv) for an 8-hour working day (comparable limits set by different countries); 240 μg m^{-3} in air are considered as nasal toxicant lowest observable level [17];
about 100 μg m^{-3} air is the tropospheric threshold conc. for effects on plants (tobacco, etc.) [18]);

- Redoxpotential in aqueous solution:

$$(O_3)_{gas} + 2H^+ + 2e^- \rightarrow (O_2)_{gas} + H_2O \quad E_0^H = 2.07 \text{ V}$$
$$(\text{for: pH = 0 and } [O_2/O_3] \text{ in gas), cf. [19].}$$

1.4
Secondary Oxidants

As outlined in Fig. 2, in water a large fraction of aqueous ozone is generally transformed into secondary oxidants and reductants such as:

- H_2O_2/HO_2^-, hydrogen peroxide, a fraction of which is present in its dissociated form (pK$_a$ 11.3);
- $HO_2^{\cdot} / O_2^{\cdot-}$, hydroperoxyl radical, that dissociates to form the superoxide anion (pK$_a$ 4.7);
- $OH^{\cdot}$, hydroxyl radical, the most important and most reactive secondary oxidant. It reacts with all types of solutes that can be oxidized, even with bicarbonate and carbonate ions;
- $HCO_3^{\cdot}/CO_3^{\cdot-}$, carbonate radical, a very selective tertiary oxidant;
- $HOBr/BrO^-$, hypobromous acid and hypobromite (in waters that contain bromide) (pK$_a$ 9);
- MnO_4^-, permanganate (in waters that contain manganese);
- $RO^{\cdot}$, oxy radicals, possibly produced from peroxy radicals, $ROO^{\cdot}$ that form when molecular oxygen adds to organic carbon-centered radicals and that transfer an O-atom to a receptor.

All pathways presented in Fig. 2 may lead to different products. Although different pathways may proceed simultaneously, only a few of them will significantly

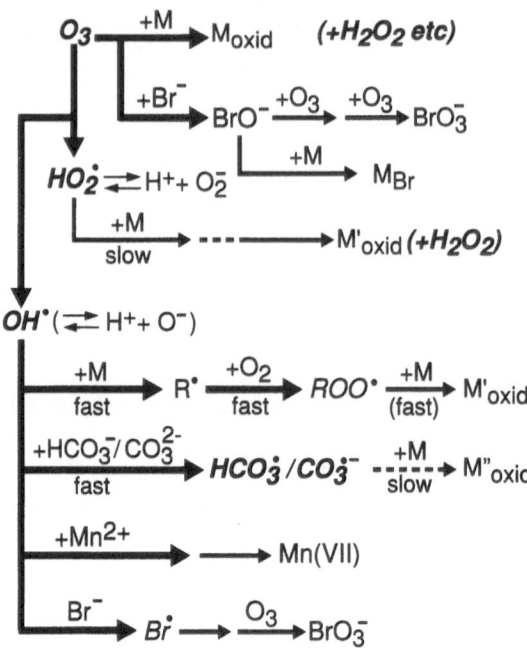

Fig. 2. Direct reactions of ozone with solutes (M or Br⁻) and reactions of dominant secondary oxidants (adapted from [20])

predominate in a given type of water. Most of them are controlled by different process paramters and by different water quality parameters. The different reactions can, however, be separated and combined to yield submodels which can be studied and tested more easily.

1.5
Some Characteristics of a Conventional Ozonation Process

A conventional unit for an ozonation process is presented in Fig. 1. For water treatment processes O_3 is typically produced on-line in an O_3 generator fed with dried air or oxygen (cf. [2, 3, 21]). Typically O_3, is formed by a silent electrical discharge (corona discharge) in an air gap between two electrodes separated by a dielectric (e.g. between glass tubes). The electrodes are connected to a high voltage generator which supplies a peak voltage in the 10 to 30 kV range. Frequencies used are in the low (about 50 Hz), the medium, or the high range (> 1000 Hz). Typically concentrations of O_3 achieved are in the range of 15 g m⁻³ when air is the feed gas, but in the range of up to 150 g m⁻³ when oxygen is applied. To transfer the O_3 into the aqueous phase, the ozone-containing gas is diffused into a counterflow of water via porous plates or turbine-type diffusers located at the bottom of contact columns (typical depth 2–3 m). For wastewater treatment packed columns are also applied. The rate of O_3 transfer may become limiting when waters with a very fast O_3 demand are treated. Although the water counterflows the rising O_3 containing gas, the offgas may still contain some residual O_3 that is generally destroyed by thermal or thermal/catalytic processes. Whenever a high degree of disinfection is required, the ozonized water passes through an additional chamber giving a retention time of 10 to 20 minutes. Systems of series of a few (5–7) chambers acting as quasi stirred reactors suffice to avoid hydraulic short circuits and to simulate the characteristics of a plug-flow reactor (cf. Sect. 6.1). In practice, such a reactor performance is sometimes also simulated by inserting a series of walls into the reaction chamber to attain a series of semi-mixed sections. At the exit of the reactor, residual aqueous O_3 is often destroyed by activated carbon or by adding some hydrogen peroxide. Both processes transform a large fraction of the O_3 residual into additional OH that still contributes to a non-selective overall oxidation process (see Sects. 4 and 5). The ozonation is sometimes a primary treatment step (pre-ozonation), although the rawwater is often pre-treated by flocculation-sedimentation or even by filtration processes. Generally the ozonation is complemented by further treatment steps such as summarized in Fig. 3 and discussed in Sect. 7. For

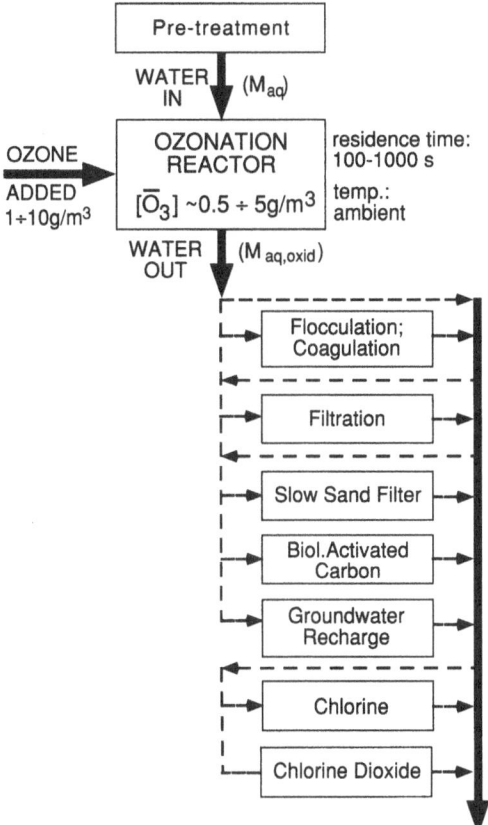

Fig. 3. Conventional ozonation of drinking water treatment including combined processes (adapted from [22])

drinking water treatment a second ozonation is sometimes installed between filtration and the microbiological process-unit (inter-ozonation).

As depicted in Fig. 3, conventional ozonation processes for drinking water treatment must be effective at ozone concentrations of only a few mg/l (< 100 µmol l^{-1}), at low ambient temperature, and within reaction times of a few minutes. In addition, O_3 is a very selective oxidant. Hence, for most applications where O_3 is used to improve waters containing a wide mixture of dissolved compounds, the advantage of ozonation is rather due to the ability of O_3 to selectively kill microorganisms and to oxidize reduced manganese and iron as well as organic compounds that contain especially ozone-reactive functional groups such as olefins, amines, sulfides, or phenolic compounds. It is because of this selectivity of ozone that low dosages often suffice to achieve the hygienic and organoleptic and chemical goals.

However, even during conventional ozonation processes, part of the O_3 is transformed into secondary oxidants such as OH radicals (cf. Fig. 2). These most reactive oxidants are predominantly consumed in non-selective but very fast reactions (µs scale). Nearly all dissolved organic compounds present in natural or contaminated waters, and even bicarbonate, act as OH˙ sinks. This makes OH˙ reactions relatively inefficient for oxidizing specified target compounds when these are present at relatively low concentrations in a natural water matrix or in a wastewater. In spite of this, there are interesting applications of OH˙ processes, whenever requirements to oxidize ozone-recalcitrant pollutants allow for a higher O_3 dosage.

2
Direct Reactions of Molecular Ozone

2.1
Types of Molecular Ozone Reactions

O_3 reacts with reactants primarily with its end-standing electrophilic O-atoms. The most often described types of reactions of aqueous O_3 are:

a) *Electron-Transfer Reactions*

$$O_2^- + O_3 \longrightarrow O_2 + O_3^- \tag{1}$$
superoxide ion   *ozonide anion*

$$HO_2^- + O_3 \longrightarrow HO_2^˙ + O_3^- \tag{2}$$
hydroperoxide anion   *hydroperoxyl*   *ozonide anion*

b) *Oxygen-Atom Transfer Reactions*

$$OH^- + O_3 \longrightarrow HO_2^- + O_2 \tag{3}$$
hydroperoxide anion

$$Fe^{2+} + O_3 \longrightarrow FeO_2^+ + O_2 \tag{4}$$
$Fe^{(IV)}$

$$NO_2^- + O_3 \longrightarrow NO_3^- + O_2 \tag{5}$$

$$Br^- + O_3 \longrightarrow BrO^- + O_2 \qquad (6a)$$

$$I^- + O_3 \longrightarrow IO^- + O_2 \qquad (6b)$$

(at high [I$^-$] follows: $IO^- + H^+ \rightarrow IOH + I^- \rightarrow I_2 + OH^-$)

c) *Ozone Addition Reactions*

Based on the Criegee mechanism the following plausible sequence of aqueous-phase reactions of an olefin can be selected for instruction:

Here, R and R' denote organic substituents or H-atoms. (For more details and critical information cf. Ref. [10]).

The ozonation of a primary amine in water can be described by the following example:

Here R is an organic substituent such as a tert-butyl group. (For more details and critical information cf. Ref. [11]).

In the case of reactions of O_3 with inorganic compounds, direct electron-transfer reaction seems to be highly exceptional. Only the electron-transfer reaction (2) has been critically tested by direct kinetic sspectroscopy. Even reaction (1) has so far not been experimentally distinguished from an O-atom transfer reaction. But, as explicitely tested by experiments, most other reactions of O_3 with inorganic species do not produce radicals that initiate the typical radical-type chain reactions (see Sect. 3.2). More typical are reactions where a primary O_3 addition leads to an intermediate that releases an O_2 molecule (in its short-lived singlet state). This results in an apparent O-atom transfer reaction (cf. Eqs. 3 to 8). For example, direct observations of product formations by kinetic spectroscopy has elucidated the reactions (3) and (4) (see also Sect. 3.2) [23, 24]. FeO_2^+ formed in reaction (4) has been observed to oxidize a further Fe(II) to form two Fe(III) [23]. O-atom transfer to nitrite leads to nitrate (Eq. 5) and that of bromide (Br^-) to hypobromite (BrO^-) (Eq. 6) (for more details on this reaction cf. Sect. 6.3).

An O_3 addition reaction followed by a fast rearrangement is typical for the ozonation of an organic olefinic compound (Eq. 7). This reaction sequence is well investigated and highly applied in organic chemistry (for a most comprehensive review cf. [10]). In aqueous systems the primary addition product, a five-ring ozonide (1,2,3-trioxalane) decomposes into a carbonyl and a carbonyl oxide. The latter hydrolyses to produce carboxylic acid. In case of reactions of O_3 with many further organic groupings, such as also compiled by Bailey in his 2nd. volume [11] the primary ozone-addition product often rearranges to emit an O_2 or CO_2 molecule and the organic product appears to be the net result of an O-atom transfer reaction (cf. Eq. 8). Also in organic chemistry, only a fews cases have been formulated as electron-transfer reactions that yield O_3^- (e.g. some cases of ozonation of some amines). However, formation of H_2O_2 has been experimentally detected and HO_2 is sometimes well assumed as an ozonation by-product (e.g. in the case of ozonation of alcohols and ethers).

The rarity of electron-transfer reactions to O_3 contrasts with the dominant mechanisms observed for other oxidants or disinfectants. For example, most known reactions of chlorine dioxide (ClO_2) or of the carbonate radical ($CO_3^{\cdot-}$) can be simply explained by electron transfer from substrates to these oxidants. Correspondingly, redox potentials directly allow for good structure-reactivity relationships for these other oxidants [24, 25] whereas reactions of O_3 follow such correlations only within highly restricted series of related compounds [26, 27].

2.2
Rate Law for Molecular Ozone Reactions

For conventional water treatment practice the overall rate of ozonation processes is controlled by the sum of the rate of the chemical reactions with individual types of solutes. O_3-mass transfer and mixing processes become the rate limiting steps only in cases where O_3 is very quickly consumed. However, even under such circumstances, the selectivity at which the different substrates present in the solution compete for O_3 still follows their relative rate constants and concentrations. Therefore, it is always crucial to search information on molecular kinetic parameters.

All primary reactions of ozone with dissolved compounds M can be formulated as bimolecular reactions:

$$O_3 + M \xrightarrow{k_M} M_{oxid} \tag{9}$$

The kinetics of the initial- and rate-controlling reaction have always been found to be first order with respect to both the concentration of O_3 at time t, $[O_3]_t$, and the concentration of the substrate compound at time t, $[M]_t$. For the rate of depletion of the concentration of a unique chemical M reacting with O_3 we can therefore write:

$$-d[M]_t/dt = k_M \cdot [M]_t^1 \cdot [O_3]_t^1 = \eta \, k_{O_3} \cdot [M]_t^1 \cdot [O_3]_t^1 \tag{10}$$

Here, k_M is a second-order rate coefficient that represents the sum of rate constants with which all reactive groups present in a molecule M reacts with O_3. η is a stoichiometric factor for the amount of O_3 that is depleted per M that is transformed. The differential rate law (Eq. 10) can, for instance, be integrated for a reactor which is closed for M (plug-flow, batch-type, or semi-batch-type reactor). The resulting time-dependency for the depletion of the relative residual concentration of a unique type of compound M of initial concentration $[M]_o$, then becomes:

$$-\ln [M]_t/[M]_o = k_M \cdot [\overline{O_3}] \cdot t \tag{11}$$

or:

$$[M]_t/[M]_o = e^{-k_M \cdot [\overline{O_3}] \cdot t} \tag{12}$$

Thus the logarithm of the relative residual concentration of M, when plotted vs the ozone exposure-value, $[O_3] \cdot t$, declines with constant slope, k_M.

To account for time-dependent concentrations of ozone, $[O_3](t)$, it is often more convenient to plot the log of the relative residual concentration of M vs the time-integrated ozone exposure-dose, $\int([O_3](t) \cdot dt)$. This corresponds to the linear, time-averaged concentration of the ozone that acts during time t:

$$\int([O_3](t) \cdot dt) \rightarrow [\overline{O_3}] \cdot t = c \cdot t \tag{13}$$

This time integrated ozone exposure dose is also denoted as the "c · t-value", where "c" denotes the time averaged concentration of the oxidant.*

---

* In some water treatment guidelines the "c · t -value" is based on the residual of an oxidant (disinfectant) such as measured at the exit of the reactor, c_e, and t is defined as the time at which the first 10% of a spiked compound have passed the reactor. Such a pragmatic parameter might be useful for characterizing a conventional (chlorine-based) disinfection process; however, it is a misleading parameter to describe processes where the conc. of ozone varies more rapidly or to compare different types of reactors or to characterize the formation of disinfection by-products.

In general, only such a correct (time-integrated) c · t-value is a useful key parameter for describing the kinetics of an ozonation process. (This key parameter also appears in the equation that describes the completely stirred tank reactor (see Sect. 6.1).)

The processes that are first order in [M] (and not zero-order) can be described by such simple equations (Eqs. 11, 13) only when considering individual compounds M or unique chemical structures within a compound of a unique k_M value. In contrast, lumped parameters, such as color or light absorption values, taste, total organic carbon content (TOC), assimilable organic carbon (AOC), concentration of total polyaromatic hydrocarbons, or organic halogen compounds (TOX) quantify the sum of a mixture of individual compounds that have different reactivities. The apparent reaction kinetics of such a mixture depend on its actual composition. This is not defined and changes as the reaction proceeds. Therefore, during the process a decline in the apparent rate-coefficient of the mixture is observed that represents an "ageing of the kinetics"; for such cases no constant rate-coefficient can be derived. The kinetics given in Eqs. (11, 12) can therefore be calibrated and applied only when they are based either on a uniform compound or on a reservoir compound that supplies the specific reactive species within a fast pre-equilibrium such as, for instance, provided by acid-base equilibria. Therefore, any process calibration should be based on the observation of a well defined specific and molecularly unique reference compound. Similarly, for calibrations of disinfection processes it has also been accepted that these must be based on a uniform species of a microorganism used as a reference and not on total counts of a (variable) mixture.

2.3
Rate Constants for Molecular Ozone Reactions

The rate constants for the aqueous phase reaction of molecular ozone with about 250 compounds can now be found in up-dated compilations [13, 14] and comprehensive overviews [20, 22, 26–28]. Many of the constants were determined when testing the rate laws and methods for measurements [27–31]. Because some reactions are only defined by additional information on the system or by the operations of measurements, or only relative to specified references, a safe application of data compilations still requires that the original literature be consulted carefully.

In addition to information from aqueous systems, rate data published for non-aqueous media can sometimes be transformed. Appropriate conversion factors, however, differ for different classes of compounds [27]. In addition, in water many relevant compounds change their chemical speciation: they dissociate into ions or become complexed by cations or chelated by other ligands, thus forming new aqueous species that exhibit very different properties [27–29]. However, constants for additional chemicals of interest can be measured easily in any laboratory equipped for basic physical-chemical measurements, provided that all precautions and tests are taken into account to avoid interference by chain reactions which lead to secondary reactions as well as to an additional decomposition of O_3 [27–31]. Rate constants for slow reactions (time range of seconds to minutes) can be measured by following the relative concentration of added O_3 solutions to solutions that contain a fair excess of the reactive substrate or its precursor reservoir, such as to keep the concentration of the reactant quasi unchanged during the reaction [27–31]. For a complementary approach, it is also possible to maintain a constant concentration of aqueous O_3, e.g. by a continuous supply from a gas stream to a well mixed "reactor", and to monitor the rate with which either the concentration of a dissolved compound decreases or products are formed. For some cases it is also appropriate to determine relative rate-constants by following the decrease of the concentration of the compound relative to a reference that has a rate constant of comparable magnitude (i.e. within a factor of < 3 to 4). For faster reactions (time range of milliseconds), also stopped-flow methods [23, 24] or through-flow techniques [32] have been applied. Very fast reactions (time range of microseconds) have been measured by producing the reactants (e.g. free radicals such as $OH^{\cdot}$ or $HO_2^{\cdot}$) by pulse radiolytic methods or by flash (UV) photolysis and following the formation of (transient) products in presence of dissolved O_3 by kinetic spectroscopy [33–39]. New data may also result as fitting prarameters when predictions from well set model calculations become compared with experimental data. However, for further applications or data-compilations only very critically tested rate-data should be used. In general, a high quality of a second-order rate coefficient is only demonstrated when the value that was directly measured or found by iterative processes has been experimentally shown not to float when the concentrations of O_3 and substrate and further solutes are varied by at least an order of magnitude.

In Figs. 4–6 a few inorganic and organic aqueous compounds are vertically arranged according to their second-order rate constants (left-hand axis). Because of experimental reasons this scale is given in k_{O3} – and not in k_M values (see Eq. (10)). However, the experience is that the stoichiometric factors, η, that control the ratio between the two types of rate constants is always in the range between 1 (for compounds such as olefins, nitrite, iodide, Fe(II), etc.) and 2 or 3 for most non-olefinc organic compounds. Such factors do not give a bias on the compressed scale used for Figs. 4–6 that extends over 10 log units. On the right-hand ordinands of Figs. 4–6, a corresponding inverse scale indicates the time constant τ ($t_{37\%}$) of the reaction, i.e. the time within which by Eqs. (11, 12) the concentration of the reactant M is reduced by a factor e (to e^{-1}, i.e. to 37%):

$$\tau = t_{37\%} = 1/(k_M \cdot [O_3]) \tag{14}$$

The corresponding ozonation time required to reduce the concentration of the indicated compound M to 50% or 10% would be $\tau \cdot \ln(2)$ ($= 0.69 \ \tau$), and $\tau \cdot \ln(10)$ ($= 2.3 \ \tau$), respectively.

For setting this time scale it has been assumed that the oxidation takes place in presence of 10 µmol l^{-1} O_3 (about 0.5 mg/l, as it is typical for conventional drinking water treatment) and that the stoichiometric yield factor was 1.0. For a ten times higher concentration of O_3 the time constant of the reaction would become ten times shorter, i.e. the scale would have to be lowered by one log unit.

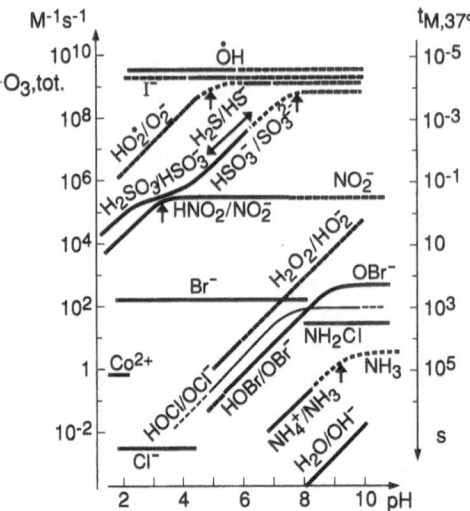

Fig. 4. Examples for aqueous second-order rate constants k_M for direct reactions of ozone with inorganic solutes as a function of pH (cf. Eq. (10)). Right-hand scale: $t_{M,37\%}$ is the reaction time (τ) required to reduce the concentration of the indicated solute by a factor e (i.e. to 37%) (cf. Eq. 14). Vertical arrows indicate pK_a values. *Assumptions:* $[O_3] = 10\ \mu M$ (0.5 mg/l); no interferences by secondary oxidants; batch-type or plug-flow reactor. Data from [28], Fig. adapted from [20]

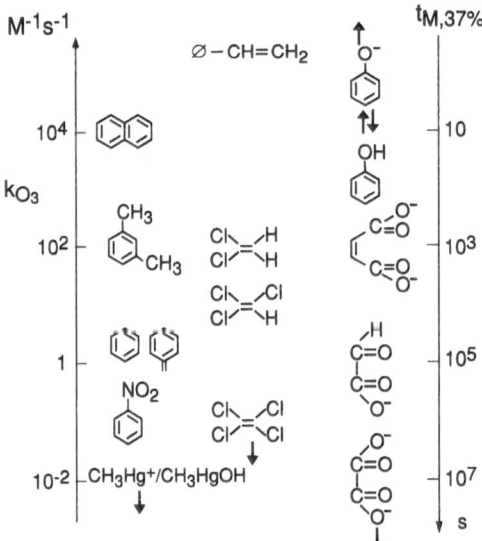

Fig. 5. Examples for aqueous second-order rate constants k_M for direct reactions of ozone with non-dissociating organic solutes (cf. Eq. (10). Right-hand scale: $t_{M,37\%}$ is the reaction time (τ) required to reduce the concentration of the indicated solute by a factor e (i.e. to 37%) (cf. Eq. 14). Assumptions as for Fig. 4. Data from [27], Fig. adapted from [20]

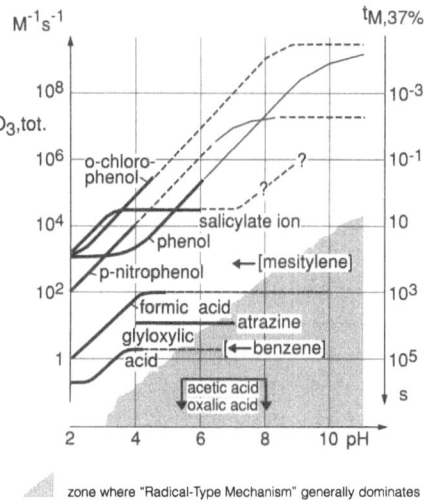

Fig. 6. Examples for aqueous second-order rate constants k_M for direct reactions of ozone with dissociating organic solutes as a function of pH (cf. Eq. (10). Right-hand scale: $t_{M,37\%}$ is the reaction time (τ) required to reduce the concentration of the indicated solute by a factor e (i.e. to 37%) (cf. Eq. 14). Assumptions as for Fig. 4. Data from [27], Fig. adapted from [20]

It has also been asssumed that the process was performed in a batch-type or plug-flow reactor (compare Sect. 6.1).

Figures 4–6 demonstrate well the high selectivity of ozonation reactions; the scales used to span selected compounds of practical interest easily covers a factor of 10^{10}. From the chemical species shown in these figures and listed in compilations we can deduce:

- **Hydroxyl radicals (OH·)** and **hydroperoxyl radicals/superoxide ($HO_2^·/O_2^-$)** react with O_3 very quickly. Their role in the radical-type chain reaction leading to an enhanced decomposition of aqueous O_3 is discussed in Sect. 3.
- **Water/hydroxide anions (H_2O/OH^-).** H_2O reacts with O_3 only when deprotonated to OH^-. Thus, this reaction becomes ten times more important when the pH is increased by 1 unit. In a reaction of a very low rate constant OH^- accepts an O-atom from O_3 (see Eq. 3). The HO_2^- produced rapidly achieves an acid-base equilibrium with its protonated form, H_2O_2 [23, 31].
- **Hydrogen peroxide/hydroperoxide anion (H_2O_2/HO_2^-)** also reacts with O_3 only in its deprotonated form (see Eq. 2). Correspondingly, the apparent rate constant based on $[H_2O_2]_{tot}$ increases tenfold per pH [31]. Both reaction products, O_2^- and $HO_2^·$, contribute to the radical-type chain reaction that enhances the decomposition of aqueous O_3 (cf. Sect. 3.2)
- **Iodide (I^-)** reacts with O_3 in a very rapid reaction to produce aqueous iodine. (cf. reaction 6b). Although the stoichiometry of the sequence of secondary reactions depends on many conditions, this reaction is sometimes applied for the analysis of high concentrations of gaseous O_3. In drinking water,

traces of oxidized iodide may lead to taste forming organic compounds. At the surface of the sea, where iodide is present at concentrations of about 10^{-7} M, I^- reduces the lifetime of deposited O_3 to about 0.01 s.

– **Hydrogen sulfide/sulfide (HS^-/S^{2-})** is generally oxidized by O_3 before O_3 is significantly consumed by any other solute. This reaction is of importance when sulfide containing reduced groundwaters are treated with O_3 (e.g. the Swiss Office of Health has approved this process even for treating sulfide-containing mineral waters. Also the EU commission proposes acceptance of this treatment).

– **Sulfite (HSO_3^-/SO_3^{2-})** also reacts immediately. Because the rate constant of SO_3^{2-} is about 10^5 times higher than that of HSO_3^-, the apparent rate constant based on $S(IV)_{tot}$ increases over the whole pH region of 2–7 by a factor of 10 per pH increment. Such reactions are of direct interest not only for the treatment of special wastewaters but also in the field of cloud chemistry where they may control the oxidation of aqueous SO_2 species into sulfurous acid (for references cf. [40]).

– **Iron(II), copper(I), manganese(II).** Fe^{+2}, at low pH, exhibits a surprisingly low rate constant [24], whereas $Cu(I)$ seems to react nearly in a diffusion controlled reaction [39]: Such reactions are of particular interest in atmospheric waters where these transition metals act as catalysts for converting atmospheric photooxidants in cloud droplets [12, 41, 42]. But reduced groundwaters are at much higher pH, and in the presence of DNOM the iron(II) may assume other speciations. In the pH-range 5.5–7.0, but in absence of humic material, Mn(II) was found to be oxidized by O_3 with a rate constant of $3 \cdot 10^3 - 2 \cdot 10^4$ $M^{-1} s^{-1}$ [42]. The oxidation processes that might lead to the formation of permanganate highly depend on the presence of DNOM. However, ozonation processes were often well applied for the removal of reduced iron and manganese such as present in reduced groundwaters [3, 43].

– **Nitrite (NO_2^-)** is oxidized quickly. This fact is relevant to improve waters when nitrite, for instance, has been produced through incomplete microbiological denitrification.

– **Ammonia (NH_3)** reacts very slowly. In addition, in the pH region below 9 a significant fraction of the ammonium nitrogen is masked by protonation as ammonium ion (NH_4^+). Hence, in the pH region of < 9, the apparent reaction-rate constant for total ammonia decreases by a factor of ten when the pH is lowered by one unit. Even in the high pH region, where most NH_4^+ deprotonates to form reactive NH_3 (e.g. pH > pK_a = 9.3), thousands of seconds of conventional ozonation are required to oxidize this compound. Experience shows that ozonation of surface waters may, however, enhance the nitrification of ammonia in succeeding microbiological processes by converting more DNOM to microbiologically degradable organic material and thus indirectly stimulate nitrifications (cf. Sect. 7.4).

– **Aqueous chlorine ($HOCl/OCl^-$)** also reacts faster at elevated pH where a high fraction of the hypochloric acid ($HOCl$) is deprotonated to form reactive hypochlorite anion (ClO^-). But even for ClO^-, the time constant for oxidation in presence of 10 µmol l^{-1} O_3 is on the order of 1000 s. This reaction has occasionally been applied to destroy an ozone residual by chlorine addition. (For more details cf. Sect. 7.1).

- **Chloramine (NH_2Cl)** is oxidized somewhat faster than aqueous chlorine, when compared in the low pH region. There is no masking of chloramine by protonation and hence there is no pH dependency. The ozonation of chloramine leads to the formation of chloride and nitrate. This reaction can be applied in eliminating chloramines in swimming-pool waters to avoid the build-up of eye irritants.
- **Bromide (Br^-)** oxidation requires a time constant of about 1000 s, when in presence of 0.5 mg/l O_3. Complementary, the lifetime of small concentrations of O_3 ($[O_3] < [Br^-]$) in waters containing 2 mg/l Br^- becomes 500 s. In marine waters, after oxidation of traces of I^- (see above), it is the Br^- (0.8 mmol l^{-1}) that will restrict the lifetime of added O_3 to less than 5 s. The hypobromite (BrO^-) formed is slowly further oxidized to bromate (cf. Sect. 6.3).
- **Alkanes, saturated alcohols, chloroalkanes, etc.** do not react with molecular O_3 at a significant rate. Rather ozone based AOPs are suited to oxidize such compounds.
- **Olefinic compounds** (e.g. oleic acids or styrene) react within seconds (see Fig. 5). However, when the H atoms in α-position to the C-C double bond are substituted by chlorine atoms, the double bonds become deactivated. Therefore, perchloroethylene or trichloroethylene, contaminants often found in ground waters of urban areas, cannot be oxidized by molecular O_3 within a reasonable reaction time. Rather ozone based AOPs are suitable for oxidizing such compounds.
- **Benzene and pyrene** react only within days. Derivatives react somewhat faster when the aromatic ring is substituted by methyl and methoxy groups, that typically activate the aromatic ring for electrophilic reactants: e.g., the rate is enhanced by a factor of 7 per methyl group substituent (c.f. **toluene, xylene, mesitylene**). A Hammett plot of the log k-values vs σ shows a very large slope ϱ [26, 27, 43].
- **Polycyclic aromatic hydrocarbons** (systems of condensed aryl rings, some of which are considered to be carcinogenic) typically react within seconds.
- **Phenols** react within seconds (cf. Fig 5). Thereby the phenoxide anion is reacting 10^6 times faster than the non-deprotonated phenolic compound. Thus, above pH 4, where at least 10^{-6} parts of typical phenols are dissociated, the rate of reaction is controlled by the small fraction of this more reactive species, and the apparent rate constant increases with the degree of dissociation (i.e. by a factor of 10 per pH unit) [27]. In the pH-range above the pK_a value of the phenolic group, the apparent rate constants of many phenols approach diffusion controlled limits. (Nitro-phenoxides seem somewhat slower.) Rate constants of non-dissociated phenols and of methoxy substituted benzenes strongly increase with the apparent σ value [25, 43].
- **Carbohydrates (sugar compounds)** react very slowly with molecular ozone [44]. They however promote very efficiently the chain reaction that accelerates the transformation of O_3 into $OH^\cdot$ and thus produce the secondary oxidant that oxidizes these compounds (cf. Sects. 3 and 4).

- **Amines (RNH$_2$, etc.) and amino acids** react quickly when the amino group is not protonated. Therefore their apparent rate constant increases tenfold per pH unit when in the pH region below the pK$_a$ of the protonated species (pK$_a$ values of amines are typically in the range of 9 to 10).
- **Pyridine, atrazine etc.** react very slowly.
- **The organic oxidation products** produced by ozonolytic cleavage of aromatic ring systems are **glyoxylate-, maleic-, oxalate-, acetate-, or formate ions** (also cf. Sect. 6.2) Of these products, only the formate ion reacts at a relatively high rate (see Fig. 5, column of compounds at right and Fig. 6). All the others will accumulate in ozonized waters as final products when no alternative pathways for oxidations, such as AOPs, become operative.
- **Dissolved Natural Organic Material (DNOM)** contain a mixture of compounds and it seems that only a small fraction of the functional groups exhibit a significant rate constant for reacting with molecular ozone; there is only a strongly limited instantaneous ozone consumption by DNOM. Most molecular structures of DNOM of aerobic surface and groundwaters appear to have rather low rate constants for consuming molecular O$_3$.

3
Decomposition of Aqueous Ozone

3.1
Lifetime of Aqueous Ozone in Different Types of Waters

Within a few seconds of addition of ozone to any natural water, part of the ozone is instantaneously consumed. Generally, this fast reaction is followed by a slower depletion of the residual ozone. For real freshwaters applied for drinking water and cooling water production, this slower decay-mode can generally be approximated by a rate law that is first order in respect to the residual O$_3$ concentration. A convenient parameter for an operational definition of this value is the second half-life, i. e., the time within which the residual concentration decreases from 50 to 25 % of its initial value. This second halflife may however still depend on the O$_3$ dosage [45, 46]. Thereby, the two values, the extend of the instantaneous O$_3$ consumption and the second half-life of dosed O$_3$, are controlled by different types of solutes and chemical reactions [45, 46]. Figure 7 gives a few examples for halflives vs pH. As a typical reference, O$_3$ dosed to raw water from mesotrophic Lake Zurich (pH 7.8, [HCO$_3^-$] = 1.2 mM, DOC about 1.2 mg l^{-1}) decomposes within 10 minutes, even after the instantaneous ozone demand of the water (about 0.1 mg l^{-1} O$_3$) has been satisfied. In groundwaters of relevant carbonate alkalinity and lower DOC or pH, the second half-life becomes higher, but when the DOC or pH is increased or when the carbonate alkalinity is decreased, it becomes shorter (see entries in Fig. 7). Practical experience resulting from measurements of the liefetime of aqueous O$_3$ in waters of varied carbonate alkalinity had already led to such information before the mechanistic chemical background has been elucidated. For example, an owner of a plumbing company who installed swimming pools and operated in various regions of Switzerland, applied in the early 1970s for a Swiss patent

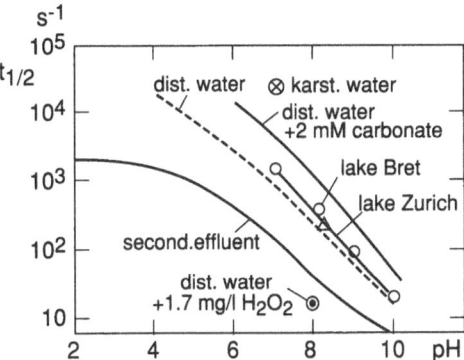

Fig. 7. Second half-life of ozone ($t_{1/2}$) dosed to different types of water as a function of pH. Within this time the concentration of ozone (about 3 mg/l at beginning of measurement) declined from 50% to 25%. For details of measurement cf. [45, 46].

– distilled water: pH adjusted with phosphate buffer (0.05 M)
– karstic groundwater: DOC =1.2 mg/l, pH adjusted by NaOH and HCl
– Lake Zurich water: DOC = 1.2 mg/l; diluted to 70%, pH adjusted by NaOH and HCl
– Lac de Bret water: DOC = 3.2 mg/l; diluted to 70%, pH adjusted by NaOH and HCl
– secondary effluent of communal wastewater plant: DOC 7 mg/l, pH adjusted by borate buffer

in which he claimed that in the case of soft water the stability of aqueous O_3 can be improved by dosing the water with bicarbonate (for references cf. [64]). (For mechanism see Sect. 3.2).

In typical groundwaters and lakewaters, the rate of the ozone decomposing chain reaction increases by about factor of 2 when the temperature is increased by 10°C [15].

3.2
Chain Reactions for the Transformation of Ozone into OH Radicals

The experience that the lifetime of O_3 in water becomes shorter when the pH or the concentration of certain types of dissolved organic compounds is increased can be rationalized when considering the chemistry of direct reactions of O_3 with solutes and a radical-type chain reaction that accelerates the further consumption of O_3: By reactions with OH^- [23, 31] or solutes or reactions on surfaces such as on AC, some O_3 is consumed and often transformed into products such as hydrogen peroxide or HO_2. Thereby the O_3 reacts apparently faster at higher pH where many solutes show a higher degree of dissociation and form anions that are the more reactive species for reacting with electrophilic O_3 (see Sect. 2.3). However, in low and moderate pH ranges, even reaction with OH^-, except in waters of highest purity, becomes rather negligible. Also the radical-type chain reactions are more easily promoted at higher pH where larger fractions of chain carriers such as HO_2^- and H_2O_2 are deprotonized [31–34].

A model that is composed of a chain of well studied individual reactions is represented in Fig. 8. It has been based on detailed direct experimental mea-

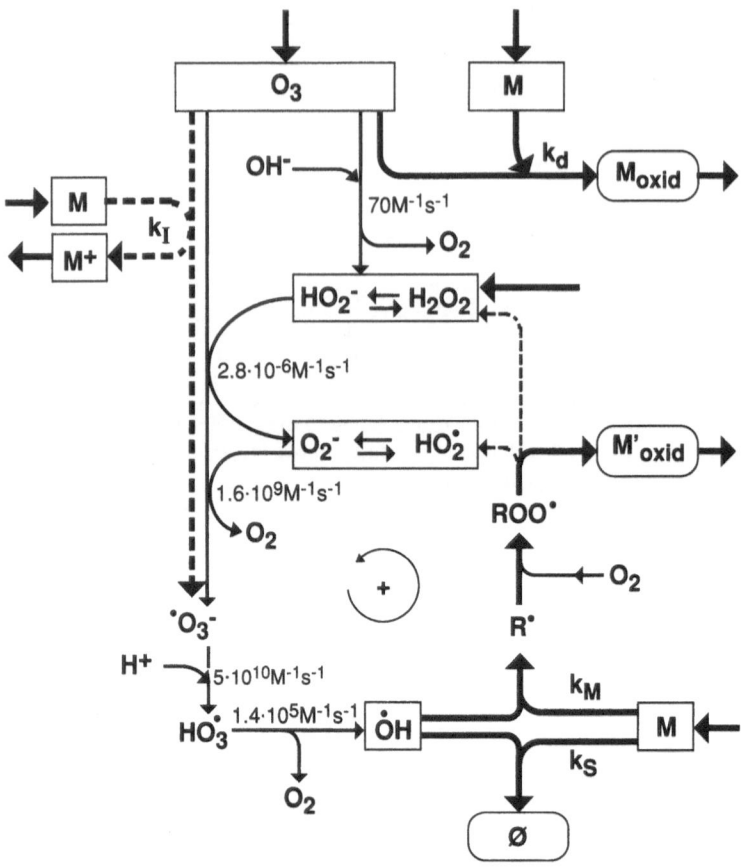

Fig. 8. Scheme of reaction of aqueous O_3. O_3 reacts either directly with a solute M (reaction-rate constant k_d) or it reacts with OH^- or M to initiate a radical-type chain reaction. Some types of M can also act as chain promoters by transforming non-selective $OH^\cdot$ into highly O_3 selective radicals (k_M), but some (e.g. bicarboante) can quench the chain reaction by just scavenging the chain carrier, $OH^\cdot$ (k_S). For further details see text. From [49], including newer data from [23]

surements of nearly all individual reactions involved and it well describes the functionalities with which different water-quality parameters effect the rate of transformation of ozone [23, 31, 33–39, 41, 44–50]. As represented by this model, the radical-type chain reactions are initiated by reactions that lead to the production of radicals that act as chain carriers. For example, O_3 transfers an O-atom to OH^- to produce HO_2^- [23]. This equilibrates to form H_2O_2 that acts as a reservoir for HO_2^-. The fraction of $[H_2O_2]_{tot}$ that is dissociated to HO_2^-, however, quickly reacts with more O_3 [31] to produce O_3^- and HO_2. The HO_2/O_2^- ($pK_a = 4.7$) thereby formed is a highly selective reactant. In ozonated (natural) waters it selectively transfers an electron to O_3 to produce additional O_3^- [33–39]. The only further significant reaction of HO_2/O_2^- in lakewater and groundwater would be dismutation or reactions with transition metals to form H_2O_2 [15]. Also this product would finally act as a chain carrier.

At very high pH O_3^- has been observed to be a long-lived transient (37). However, pulse radiolytic measurements with kinetic spectroscopy indicated that, when not in extremely high pH regions, pH buffers (or OH^-) quickly protonate O_3^- to form HO_3 (pK_a (HO_3) = 6.1). This seems either to decompose within 6 μs to form $OH^\bullet$, or to directly produce $OH^\bullet$ [32, 33, 37, 39] (see Fig. 8). HO_3/O_3^- itself is considered to be a rather weak reactant. The chain of reactions is then continued as follows: In pure water O_3 itself can become a main scavenger for $OH^\bullet$ [33, 35]. Some authors assume that $OH^\bullet$ forms with O_3 a complex, $HO_4^\bullet$, that appears as an intermediate (showing a characteristic UV absorbance), but that again equilibrates with $OH^\bullet$ and O_3 and finally decomposes to reproduce HO_2 [32–34, 39]. Thereby the chain carrier, HO_2/O_2^-, is reproduced.

Most real waters that are ozonated contain other solutes, M, also designed as pollutants P, that efficiently compete with O_3 for reacting with $OH^\bullet$ (for rate constants see Sect. 5.2). Most of the reaction products become further oxidized by O_2 to form peroxy and oxy radicals that react further with O_2 and release HO_2/O_2^- or hydrogen peroxide (cf. Fig. 9 and 10). Such transformations predominate when M (P) is, for example, formate, an alcohol, or a sugar (cf. Fig. 1 in [52] and, for a comprehensive review, [48]). This means that also many types of organic solutes transform non-selective $OH^\bullet$ into the highly ozone-selective HO_2/O_2^- or H_2O_2 and thereby reproduce chain carriers. This promotes the radical type chain reaction that enhances the transformation of O_3. In contrast, other solutes such as HCO_3^-/CO_3^{2-} and some types of organic compounds, such as alkanes, alkylalcohols or alkyl carboxylic acids, scavenge $OH^\bullet$ also to produce intermediates which do not release a radical-type chain-carrier. Such solutes therefore quench the radical-type chain reaction and, thus, somewhat stabilize the aqueous ozone. (cf. e.g. [38, 41, 47–49]).

At very low pH, and in absence of O_2, also a unimolecular dissociation reaction, $O_3 \rightarrow O_2 + O(^3P)$, has been observed [47]. However, this dissociation proceeds much too slowly (k = $3 \cdot 10^{-6}$ s^{-1}) to contribute to the apparent depletion of O_3, and in presence of O_2 this reaction becomes reversible.

We may conclude: the decomposition of aqueous ozone is enhanced at higher pH or in presence of elevated concentrations of dissolved organic compounds (humic material), but it becomes slower when the pH is decreased, when the carbonate alkalinity is increased, or when such types of organic compounds scavenge $OH^\bullet$ that do not form HO_2/O_2^- or H_2O_2 during secondary reactions. Because of the complexity of the chain of reactions that proceed during ozonation of real waters and due to the chemical complexity of the DNOM that is present in all natural waters, it is advisable to experimentally determine the rate of decomposition of O_3 whenever an ozonation process is to be applied [46].

Although some quite successful applications of such models to calculate the overall kinetics of the decomposition of O_3 and formation of reactive intermediates have been published, cf. e.g. [51], quantitative predictions that are just based on this model have only hypothetical character: Many uncertainties are due to the fact that the model has rather been formulated to just account for all known reactions. In real waters there might also further reactions be involved.

In addition, accurate rate constants and pK_a values that are compiled for the individual reactions of the chain carriers are just defined by the operation of the measurement (even the relationship between pH and [OH$^-$] is rather defined by the operation of its measurement or by the composition of the solution). Also the chemistry and reactivity of DNOM varies greatly according to its origin and with ageing processes and can, therefore, not be represented by generalized reaction constants. Therefore, any model calculations or extrapolations based on compilations of reaction-rate data or on experiences from other types of waters only yield a first estimate but they can help in planning good experiments and for modelling sensitivity analysis.

4
Initiation of Ozone-Based Advanced Oxidation Processes (AOPs) by Hydrogen Peroxide or UV Irradiation

Processes based on the formation of highly reactive OH$^•$ to act as an oxidant have been designated as "Advanced Oxidation Processes" or "AOPs". More recently this term has also been applied to include water treatment technologies that are mediated by further highly reactive oxidants, or even to include homogeneous and heterogeneous UV processes.

In addition to all these different AOPs, the O_3 based production of OH$^•$ still seems to be the most economic. The process that transforms O_3 into OH$^•$ is generally initiated by elevating the pH, or by adding hydrogen peroxide ("Hydrozone Process"), or by UV irradiation. Also presence of suspensions of activated carbon might be able to stimulate such a process ("Carbozone Process"). All these initiations are generally followed by a radical-type chain reaction that proceeds in the bulk of the solution and enhances the further transformation of O_3 into OH$^•$ (see Sect. 3.2). A scheme that outlines the similarities of different O_3 based processes (but not including AC-initiation) is presented in Fig. 9. As a common key-reaction for both, direct dosage of H_2O_2 or UV irradiation, we must consider that the deprotonated H_2O_2 (i.e. HO_2^-) transfers, in a fast reaction ($k = 3 \cdot 10^6$ M^{-1}s^{-1}), an electron to O_3 to produce O_3^- and HO_2/O_2^- that act as chain carriers for the succeeding radical-type chain reaction:

$$\begin{array}{l} H_2O_2 \\ \downarrow \uparrow \\ HO_2^- \xrightarrow{\ O_3\ } \quad HO_2^• + O_3^- \end{array} \qquad (15)$$

Although only a small fraction of the total H_2O_2 is deprotonated to its reactive species ($pK_a(H_2O_2) = 11.7$); both, the fractions of H_2O and of H_2O_2 that are deprotonated and thereby become reactive to react with O_3 increase just tenfold when the pH is increased by 1 unit [31]. In the whole pH range below pH 11.7, and whenever the concentration of H_2O_2/HO_2^- is kept above 0.1 µmol l^{-1}, H_2O_2/HO_2^- becomes a more important reactant for O_3 than the hydroxide anion (H_2O/HO^-).

To enhance the transformation of O_3, H_2O_2 is directly dosed to the ozonated system. But also a transformation of O_3 itself into H_2O_2 may act as the source for H_2O_2. To accelerate this, the pH is increased or, more effectively, UV photolyses

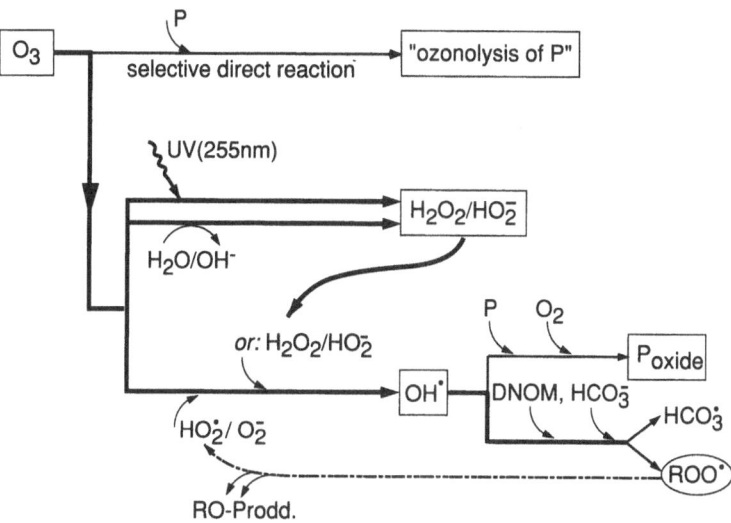

Fig. 9. Pathways for transformation of O_3 by reactions with pollutants (P), OH^-, H_2O_2, or by UV photolysis. Reactions with OH^- as well as UV photolyses produce primarily H_2O_2. In natural waters the initiated chain reaction is promoted by some reactions with DNOM and inhibited, e.g. by some other reactions with DNOM and by HCO_3^- (adapted and updated from [22]). Not shown is the initiation by activated carbon (cf. Sect. 7.5)

of O_3 is applied [52, 54, 56]: If the aqueous O_3 is exposed to UV light of a wavelength $\lambda < 300$ nm it is very efficiently photolysed to form an excited O atom $(O(^1D))$ which, in water, adds to H_2O to form H_2O_2 [50]:

$$O_3 \xrightarrow{\text{UV}} O_2 + O\,(^1D) \xrightarrow{H_2O} H_2O_2 \tag{16}$$

This aqueous phase reaction is very different from the corresponding photolytic reaction that occurs in the gaseous phase where the $O(^1D)$ formed from UV photolyzed ozone reacts with gaseous H_2O to directly produce two $OH^\bullet$.

To stimulate AOPs in drinkingwater by the H_2O_2/O_3 process, a H_2O_2 dosage of about 0.3–0.4 mg/l H_2O_2 per mg/l of O_3 has sometimes been recommended. In general an optimal H_2O_2 dosage strongly depends on the chemical composition of the water in respect to promote the subsequent chain reaction. For the photolyses of ozone the low pressure mercury lamp, as commercially used for UV disinfection, is the most frequently applied UV source. It has a strong emission at 255 nm. When penetration depths of the order of 5 cm are considered, such UV still penetrates most natural waters easily; the wavelength at 255 nm is just above the absorption bands of nitrate or carbonate, but it ideally corresponds with the short UV absorption band of aqueous ozone ($\varepsilon_{254nm} = 3150\ M^{-1}cm^{-1}$) [15]. It should be noted that in this UV range H_2O_2 is photolysed (to form 2 $OH^\bullet$) about 50 times slower than O_3. A UV/H_2O_2 based AOP (in absence of O_3) therefore requires a relatively high dosage of H_2O_2 and (or) a much longer UV-exposure-time than the UV/O_3 process.

As indicated in Fig. 9 (and described in Sect. 3.2), the initiated radical-type chain reaction is then promoted by O_3 itself or by some types of organic impurities. In most waters the kinetic chain lengths of such chain reactions finally dominate the transformation of O_3. Therefore, about the same yield of $OH^{\cdot}/\Delta O_3$ seems to be achieved when either OH^- (elevated pH), UV, directly added H_2O_2 or suspensions of AC (cf. Sect. 7.5) act as an initiator, and the chemical effect of all these ozone-based AOPs are comparable (also see Ref. [53]). Only at relatively high concentrations of H_2O_2 does H_2O_2 itself significantly scavenge and reduce $OH^{\cdot}$ (for comparison of scavenger rates see Sect. 5.3, Fig. 14 or Ref. [53]).

In most waters about half of the dosed ozone can be transformed into $OH^{\cdot}$. The other half seems to become consumed by oxidized intermediates. Although this stoichiometric relationship depends on various conditions, it well approximates most experimental results on different types of ozonation processes and ozone based AOPs in natural waters and wastewaters (see Sect. 5).

A combined O_3-UV process performed at low pH or high UV intensities, i.e. at conditions where the formation of H_2O_2 is much faster than its subsequent reactions with O_3, results in the formation of residual H_2O_2 [57]. This by-product can act as a reductant or oxidant and, therefore, interfere with subsequent chlorination or ozonation processes. Residual H_2O_2 also has to be avoided in ultra pure waters as used in the pharmaceutical industry.

A comprehensive review on the development and application of O_3 based AOPs has been included in a publication by Peyton [55]. Further references can be found in many papers that describe applications of such processes [52, 54, 56, 58, 59]. For the oxidation of O_3 and UV recalcitrant trace-impurities, such as for the oxidation of aqueous chlorinated solvents, comparable effects seem to be achieved per formed $OH^{\cdot}$ when this is produced from UV photolysed H_2O_2 or from any type of O_3 based AOPs [15, 41, 52, 53, 58–60].

5
OH Radical Reactions and Reactions of Secondary Oxidants

To qualify and quantify the performance of an ozonation process, it is important to distinguish between oxidations that are due to the direct reaction of molecular O_3 and those that are due to reactions of $OH^{\cdot}$ or further oxidants: These various oxidants are controlled by different process parameters and by different water qualities. They generally exhibit different selectivities and lead to very different product formations. For instructive examples see e.g. [20, 60–62]. To separate and access the effect of O_3 and secondary (and tertiary) oxidants, the time-integrated concentration of all oxidants present during ozonation must be accounted for. Thereby, for many processes molecular O_3 and $OH^{\cdot}$ are the key reactants that kinetically control all the dominant reaction sequences.

5.1
Types of OH Radical Reactions

$OH^{\cdot}$ can oxidize dissolved inorganic or organic compounds by different types of reactions (see Fig. 10): It can abstract an electron from a compound such as

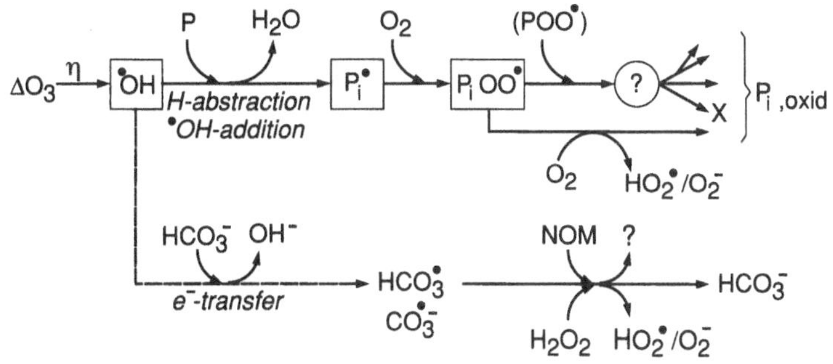

Fig. 10. Reactions of OH$^\bullet$ with an organic micropollutant P leading to a great diversity of oxidized compounds. Adapted from [53]

from a bicarbonate or carbonate ion and become reduced to OH$^-$ (*electron transfer reactions*). Alternatively, it can abstract an H-atom from many different sites of an organic molecule to yield H_2O and a substrate radical (*H-atom transfer reaction*). It adds also to the double-bond system of olefinic or aromatic hydrocarbons to form a C-centered radical with a hydroxyl group at the α-C atom (*OH$^\bullet$ addition reactions*). Therefore, by such reactions C-atom localized organic radicals are formed which may quickly add an oxygen molecule (a biradical-type species) to form reactive peroxy radicals that may become transformed into oxy radicals. Further reactions then often lead to an abtstraction of an H atom by dissolved oxygen and the formation of hydroperoxyl radicals, hydrogen peroxide, and series of peroxyls and peroxides, aldehydes, acids, etc. During the process different types of radical species are simultaneously present. Thereby (most) sequences of reactions of radicals are finally only terminated by radical-radical combinations and disproportionations. Such reactions between a large variety of different radicals even increases the large amount of different types of product formations and its dependency on many reaction parameters. Our understanding is therefore based just on a few selected examples of well performed detailed chemical and reaction kinetic studies [48]. References for further information can also be found from original research studies that are cited in compilations of rate data [13, 14]. Most of the exemples are, however, from experiments performed with pulse radiolysis or flash photolysis, where the intermediate radicals are present at much higher steady-state concentrations than during ozonation-based processes. There radical-radical reactions become faster and the radicals have less time to undergo unimolecular transformations or to react with other substrates.

5.2
Rate-Law and Rate-Constants for OH Radical Reactions

In contrast to the restricted knowledge on the complex product formations, good direct information is available for the kinetics with which OH$^\bullet$ reacts initially with all types of substrate compounds (see Fig. 10). For reaction kinetics we can formulate the reaction between OH· and a substrate molecule M (or pollutant P) by:

$$OH^· + M \xrightarrow{k_M} M_{oxid} \tag{17}$$

All primary OH· reactions are first order in both the concentration of OH· and the concentration of the substrate molecule. Their rate law can therefore be presented as:

$$-d\,[M]_t/dt = k_{OH^·,M}\,[M]_t^1\,[OH^·]_t^1 \tag{18}$$

where $k_{OH^·,M}$ is the second-order reaction-rate constant for OH· to react with a substrate compound M. Eq. (18) integrates for a reaction time (t), in case of a batch-type or plug-flow reactor, to:

$$-\ln\,[M]_t/[M]_O = k_M \cdot [OH^·]_{ss} \cdot t \tag{19}$$

or:

$$[M]_t/[M]_O = e^{-\,(k_M \cdot [OH^·]_{ss} \cdot t)} \tag{20}$$

Thereby the $[OH^·]_{ss} \cdot t$ – term can also be considered as a "c · t – value" in which c denotes the concentration of OH·. Since the early 1960s, reaction-rate data and mechanisms have been studied and compiled for hundreds of such reactions [13, 14]. Many of these measurements were performed by radiation chemists; OH· is a main reactant produced during water radiolysis. Therefore, kinetic data have been a basic tool for interpreting irradiation effects in nuclear reactor coolants, or when radiation is applied for cancer treatment, or food preservation or for initiating radical type (emulsion) polymerizations, etc. Some additional relative data for pesticides of direct concern for treatment of drinking water or leakage waters have also been deduced from concurrent reactions of ozonation based OH· formations [63]. Such rate constants for reactions of OH· now present key information for oxidations initiated by all OH· based AOPs. Some attention should be paid to the fact that discrepancies between "best" rate constants listed in compilations for a same compound but measured by different methods can easily vary by a factor of two. Only through a critical evaluation of the original literature can one find data sets that have been measured relative to a same reference, or that are based on a comparable operation. Often it seems even useful to recalibrate appropriate standards by one's own experimental comparisons with series of well known references (see Sect. 5.3). As reaction rates of OH· in the condensed phase often approach diffusion controlled limits, their selectivities are much compressed when compared with those measured for gas-phase systems and neither gas-phase rate-data nor gas-phase product formations can easily be converted to aqueous-phase data.

Some of the rate constants are compared in Fig. 11 where different target molecules are arranged according to the rate constants given on the left ordinate. When comparing this highly restricted scale with that used for presenting

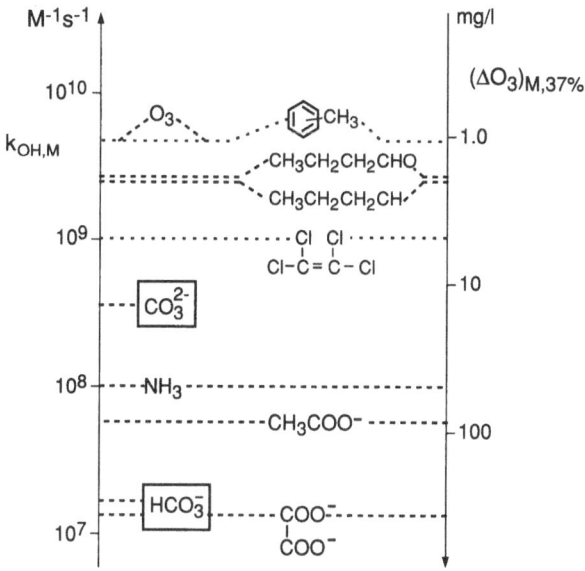

Fig. 11. Examples for aqueous second-order rate constants $k_{OH,M}$ for reactions of OH⋅ with different solutes M (cf. Eq. (18). Right-hand scale: $(\Delta O_3)_{M,37\%}$ is the dosage of ozone to be transformed into OH⋅ to decrease the concentration of the indicated substrate compound M to 37% of its primary concentration. This scale is calibrated for ozonation of a eutrophic lakewater (DOC = 4 mg/l), $[HCO_3^-]$ = 1.6 mM, pH = 8.3). *Assumptions:* η = 0.5; batch-type or plug-flow reactor. Adapted from [20]

the range of direct ozonation reactions (Fig. 4 to 6) it is evident that OH⋅ reactions are much less selective than the direct reactions of O_3. In addition, we learn from the entries in Fig. 11 and some further data that the reacting groups of a molecule contribute in a rather additive way to the overall rate constant of a molecule: The sum of the rate constants for OH⋅ to abstract an H-atom becomes the higher the more C-H bonds a molecule presents. Thereby, methyl-group H-atoms contribute a smaller fractional rate constant than those present in methylene groups or in a benzene ring-system, or those that are bond to a tertiary C atoms. In addition, OH⋅ adds to olefinic double bonds or to aromatic ring systems with a comparable rate constant as it abstracts an H atom from a methylene group. For different series of compounds that means:

- **Organic compounds of medium to high molecular mass** react quickly. Even with organic molecules of medium molecular mass, OH⋅ reacts in nearly diffusion controlled reactions. Thus, most aqueous rate constants for organic molecules are in the region between 10^9 to $10^{10}\,M^{-1}s^{-1}$. (The narrow range within which most rate data for organic substrates are found allows one to select an adequate reference compound for performing process calibrations).
- **Organic compounds of low molecular mass or of high degree of oxidation.** Methyl group H-atoms, when not activated by especial adjacent substituents, become abstracted by OH⋅ in a relatively slower reaction and the

electron transfer reaction from carboxylate anions to OH• also occurs rather slowly. Accordingly, acetate or oxalate exhibit a rate constant that is one and two orders of magnitude below that of medium sized organic molecules.

- **Halogenated organic molecules.** Organic bound halogen atoms do not react with OH• and the rate of addition of OH• to an olefinic C-C double bond becomes decreased when the H-atoms in alpha position to the double bond are substituted by halogens. Thus, (i) OH• cannot react with non-olefinic perhalogenated compounds such as CFCs, including carbon tetrachloride; (ii) H-CFCs, including chloroform and bromoform and chlorinated C_1 to C_3 alkanes and tetrachloroethylene (used as solvents), have rather low rate constants for reacting with OH•.
- **Bicarbonate and carbonate ions** transfer an electron to OH• in a reaction of rather low rate constant. However, in fresh waters, these ions are present at relatively high concentrations and they therefore often become the dominant scavengers for OH• (see later). Because H_2CO_3/CO_2 does not react, and because CO_3^{2-} reacts 30 times faster than HCO_3^-, the scavenging effect of the carbonate system increases with pH [45, 64].
- **Sulfate, phosphate, nitrate, chloride, etc.** are extremely slow in reacting with OH•; their OH• scavenging effect can generally be neglected.
- **Bromide, hypobromous acid, and hypobromate** react with OH• with a significant rate constant (cf. Sect. 6.3).

5.3
Competition for OH Radical Reactions

Figure 12 summarizes the reaction system in which OH•, that is produced from transformed O_3 (yield factor η), reacts with a specified micro-pollutant M or probe compound P (second-order rate constant $k_{OH,M}$ or, respectively, $k_{OH,P}$), and is concomitantly consumed by the sum of all solutes that act as OH• scavengers S (second-order rate constant $k_{OH,S}$). Kinetic evaluations have been given for this and similar systems by a unified formulation [cf. e.g. 20, 45, 65, 154]. One of the results formulated for a water that contains i types of scavengers S_i (including O_3 and M and P), and that is treated in a plug-flow or batch-type reactor, is:

$$-\ln ([P]_t/[P]_0) = \eta \cdot (\Delta O_3) \cdot k_{OH,P}/\Sigma (k_{OH,S} \cdot [S_i]) \qquad (21)$$

where t_{OH}, the mean lifetime of an OH•, is:

$$\tau_{OH} = 1/(\Sigma(k_{OH,S} \cdot [S_i])) \qquad (22)$$

Equation 21 shows that the logarithm of the residual concentration of P declines linearly with the amount of O_3 that is transformed into OH•, that is with $\eta \cdot (\Delta O_3)$ (also compare below and Fig. 13). The (negative) slope of the line increases with the rate constant $k_{OH,M}$ (respectively with $k_{OH,P}$), relative to the pseudo first-order rate-constant with which OH• becomes consumed by the sum of all OH• consuming reactants present in a water, $\Sigma(k_{OH,S} \cdot [S_i])$. From this,

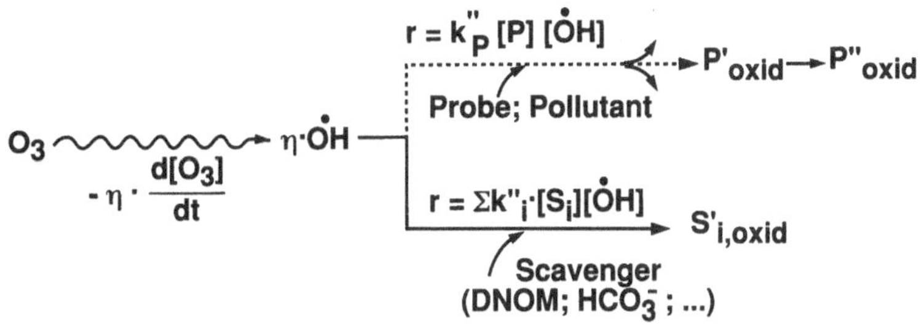

$$r = k''_P [P] [\overset{\bullet}{O}H]$$

P'$_{oxid}$ → P"$_{oxid}$

Probe; Pollutant

$$O_3 \sim\sim\sim\sim \eta \cdot \overset{\bullet}{O}H$$
$$-\eta \cdot \frac{d[O_3]}{dt}$$

$$r = \Sigma k''_i \cdot [S_i][\overset{\bullet}{O}H]$$

S'$_{i,oxid}$

Scavenger
$$(DNOM; HCO_3^- ; ...)$$

Fig. 12. Scheme of competition for reaction with OH radicals. A trace compound P (probe or pollutant) is highly protected by the sum of further OH radical scavengers S

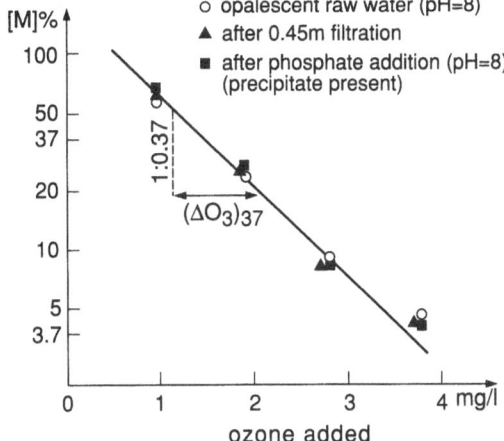

Fig. 13. Elimination of an ozone-resistant organic micropollutant M (benzene) which was spiked as a probe to samples of water from eutrophic Lake Greifensee (DOC = 3.6 mg/l; $[HCO_3^-] = 1.9$ mM; pH = 8). The reaction has been shown to be only controlled by OH radical reactions [45]. (For large series of comparable measurements also see references given in [53])

the required amount of ΔO_3 to be converted into OH$^\bullet$ to decrease the concentration of a trace compound P by a factor of e (to 37%) becomes:

$$(\Delta O_3)_{37\%,P} = (\Sigma (k_{OH,S} \cdot [S_i]))/(\eta \cdot k_{OH,P}) \qquad (23)$$

The corresponding dose of ΔO_3 required to decrease the concentration of P to 50% or 10% would be ln(2) (= 0.69), and ln(10) (= 2.3), respectively, times this $(\Delta O_3)_{37\%}$ value*).

---

* In earlier publications the Ω-value has been applied to designate $(\Delta O_3)_{37\%}$. The replacement of this has become recommendable; $(\Delta O_3)_{37\%}$ has allowed to apply a consistent notation for further OH$^\bullet$ precursors A and to compare with such other AOPs by introducing a general expressions, $(\Delta A)_{37\%,P}$ (for lit. review cf. [53]).

Figure 13 presents a typical example of the elimination of an O_3 resistant molecular species M (a micropollutant P) that only reacts with OH$^\cdot$. The probe, in this case benzene, was added in trace amounts to a sample of water from a eutrophic lakewater. Then different doses of O_3 were added and the benzene residual was measured after all O_3 was depleted. The result shows, that, once an instantaneous consumption of O_3 is met, the log of the relative concentration of P declines linearly with the O_3 dosage. By such measurements Eqs. 21 and 23 have been critically tested and calibrated for more than 50 different types of waters and for varied conditions and different probe compounds P. Also different OH$^\cdot$ precursors, A, have been inter-compared as they are applied for other types of AOPs or as they occur also in photolytic processes that produce OH$^\cdot$ (for references see [53]). For such comparisons the UV photolysis of H_2O_2 is a most convenient and "clean" OH$^\cdot$ source that, in absence of bicarbonate, can help as a reference to calibrate different systems. (In presence of carbonate, OH$^\cdot$ may produce carbonate radicals that consume some H_2O_2). Many experimentalist's experiences based on different projects that were performed over the last two decades show that such simple measurements such as represented in Fig. 13 can effectively be used to calibrate most types of waters with respect to the effect of OH$^\cdot$ reactions. Because very high OH$^\cdot$ dosages are experimentally difficult to achieve without changing conditions, some waters must be diluted before calibration; the response has however always been shown to change linearly with dilution. Basically, this method for calibration is not different from that applied by atmospheric chemists who have estimated the global mean tropospheric OH$^\cdot$ concentration by observing the rate of depletion of ubiquitous 1,1,1-trichloroethane, that represents an O_3 resistant and sunlight fast probe for OH$^\cdot$, and that is emitted into the global atmosphere with well known annual rates.

From the probe compounds that have been critically evaluated for aqueous systems, quite a few can be recommended for applications in ozone chemistry [45, 53]. For example, 1-chlorobutane is one of the outstanding probes that can be easily determined by GC (internal GC standard: 1-chloropentane). The advantage to be a less volatile probe, however, has 4-chlorobenzoate that can be determined at low concentrations by HPLC [53, 66]. Even some ozone resistant OH$^\cdot$ probes that already occur in the rawwaters to be treated have been thoroughly tested (e. g. toluene in lakewater [67], or atrazine in some drinking waters [68]).

Such probe-compound based calibrations of water qualities considering OH$^\cdot$ processes allows one to predict the rate of elimination for any other O_3 and carbonate radical resistant micropollutant, provided its relative reaction-rate constant for reaction with OH$^\cdot$ is sufficiently known to be related with that of the reference. As calibrations of waters are all based on processes that are first order with respect to the concentration of the probe (Eq. 21), this probe compounds must again consist of a unique chemical species (see arguments given in Sect. 2.2. for the calibration of O_3 reactions).

Figure 11 demonstrates the relationship between the amount of transformed O_3 that is required for the oxidation of a compound by the OH$^\cdot$ mechanism and the appropriate $k_{OH,M}$ values. For this, the right-hand scale represents a log scale for $(\Delta O_3)_{37\%}$ values that corresponds with the log scale of rate constants given at left. $(\Delta O_3)_{37\%}$ is scaled for the treatment of a representative eutrophic lake-water. In this particular water DNOM is the dominant scavenger for OH$^{\cdot}$. To treat this water, only about 1 mg/l of ozone had to be dosed and transformed to decrease the concentration of O_3 resistant organic micropollutants of medium molecular weight to 37%. Correspondingly, in the case that a water would contain twice the concentration of dissolved organic material and alkalinity as lakewater Zurich, the required amount of $(\Delta O_3)_{37\%}$ would become twice times that indicated in Fig. 11; the $(\Delta O_3)_{37\%}$ – scale would have to be lowered by a factor of 2 (i.e. 0.3 log units). In case of rawwaters used for drinking water production (TOC < 5 mg/l), about 10 mg/l of $(\Delta O_3)_{37\%}$ generally provide enough OH$^{\cdot}$ to eliminate even those organic compounds that only contain highly oxidized or halogenated carbon groupings. Exceptions are perhalogenated carbon compounds such as CFCs.

In Fig. 14, the top abscissa is scaled for the pseudo first-order rate-constant (in s^{-1}) with which OH$^{\cdot}$ is consumed. This value is reciprocal to the OH$^{\cdot}$ lifetime, τ (compare Eq. (22)). Compared with this scale are a few intercalibrated scales that show the concentrations of different common types of solutes S for which the $k_{OH,S} \cdot$ [S]-value yields a comparable first-order rate-constant for consuming OH$^{\cdot}$. We can estimate the effect of a sum of different scavengers by adding the vectors we read on the appropriate abscissa. When we, for instance, sum the vectors for the relevant ubiquitous solutes present in a drinking water of acceptable quality (S = bicarbonate + carbonate ion + DNOM) we find for the consumption of OH$^{\cdot}$ a rate constant in the order of 10^5 s^{-1}. This gives the OH$^{\cdot}$ a mean lifetime in the order of 10 micro seconds.

1-Octanol, represented on the second abscissa of Fig. 14, has often been used as a primary reference-scavenger and as an experimental surrogate for DNOM; this compound exhibits the same speciation in all waters (i.e. it is independent of pH, etc.) and its absolute rate constant for reacting with OH$^{\cdot}$ has been well established and listed in compilations. The effects of different types of DNOM and other ubiquitous solutes that occur in freshwaters, wastewaters and cloud-waters can be compared with this scale. E.g., as shown by the 3rd abscissa, the weight-based coefficient for DNOM-DOC of freshwaters (and secondary effluents of communal wastewater treatment-plants) is only about half of that of 1-octanol. Here the scale for DNOM-DOC has been based on the average that resulted from more than 20 different Swiss surface waters and samples from a reservoir in southern US (Chapel Hill) and secondary effluents of communal wastewaters (cf. [45]). Also many additional data we have determined within more recent research contracts are within the ranges we had determined before. However, for DNOM of a series of further US drinking waters another research group has determined a somewhat different coefficient [69]. Such discrepancies might well be due to different operations of measurements.

Calibrating waters in respect to rates of OH$^{\cdot}$ scavenging gives constant calibrations only as long as the reactivity of the solutes for consuming OH$^{\cdot}$ remains constant (cf. Eqs. 21, 22}. Generally, this condition is met when medium or large sized organic molecules control the scavenging. In that case, the sum of the oxidized products exhibits a rate constant for OH$^{\cdot}$ scavenging that is comparable

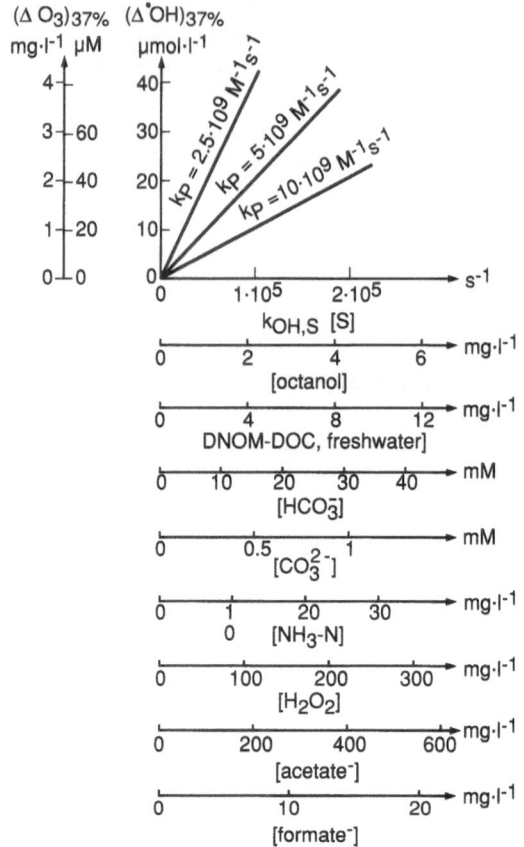

Fig. 14. $(\Delta O_3)_{37\%}$ and corresponding $(OH^\bullet)_{37\%}$ values required to eliminate pollutants P that react with $OH^\bullet$ with different reaction-rate constants to 37% of the preceeding concentration. The values are plotted vs the rate constant with which $OH^\bullet$ become consumed by $OH^\bullet$ scavengers. The series of abscissa compare the concentration effects of different types of scavenging solutes. Extracted from Fig. 1, Ref. [53]

to that of the parent compound. Hence $\Sigma (k_{OH,S} \cdot [S_i])$ remains rather constant, even when a fraction of the molecular structures of a scavenger has already been oxidized. This empirical observation can be supported by considering an example: When n-octanol is oxidized, many oxidation products accumulate. But the sum of all these, by artefact, exhibits a scavenging rate about the same as the parent molecule. For instance, the sum of the rate constants of four ethanol or two butanol species is the same as that of one 1-octanol. Also systems that contain slowly reacting scavengers, such as bicarbonate, do not kinetically age during ozonation: such scavengers only contribute a significant effect when present at quite high concentrations and these are not significantly changed during observation. The scavenging effect of small-sized organic molecules (such as methanol or formaldehyde) must, however, be tested very carefully;

during oxidation the concentration of such molecules changes, and the oxidation products have different reactivities.

Particulate matter, such as present in eutrophic lakewaters, does not act as a significant scavenger; the same rates for the elimination of OH$^\bullet$-probe molecules have been found prior to and after filtration of different lakewaters that contained high concentrations of autochthonous (plankton derived) particles. An example is included in Fig. 13.

The ordinate of Fig. 14 shows the amount of OH$^\bullet$ required to reduce the concentration of an O_3 resistant trace compound by a factor of e (i.e. to 37%). Different lines have been drawn to represent trace pollutants of different $k_{OH,P}$-values (compare with Fig. 11). The linear relationships indicate that the dose of OH$^\bullet$ required to achieve a specified degree of oxidation of a trace compound increases linearly with the pseudo first-order rate-coefficient with which the OH$^\bullet$ becomes consumed by the OH$^\bullet$ scavenging solutes (also cf. Eq. (23)).

Figure 15A gives information on the steady state concentration of OH$^\bullet$ achieved during an ozonation process when 2.4 mg/l of O_3 are transformed per 100 s into OH$^\bullet$ and when the DNOM-DOC consumes OH$^\bullet$ with a rate constant such as that indicated in Fig. 14.

5.4
Role of Carbonate Radicals ($HCO_3^\bullet/CO_3^{\bullet-}$)

In presence of bicarbonate and carbonate ions a fraction of OH$^\bullet$ formed during a process will produce some carbonate radical ($HCO_3^\bullet/CO_3^{\bullet-}$, $pK_a = 7-9.6$) [14]. This acts as a very selective further oxidant. For discussion of representative reactions cf. e.g. Ref. [14, 25].

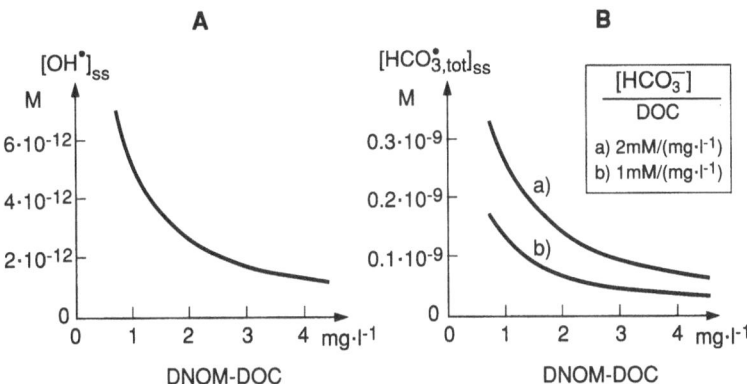

Fig. 15. Steady-state concentrations of OH$^\bullet$ and $HCO_3^\bullet/CO_3^{\bullet-}$ that occur in a water as a function of DOC.
Assumptions: **A:** 2.4 mg/l O_3 are transformed within 100 s (i.e. 0.5 µM $\cdot$ s^{-1}) to produce OH$^\bullet$ with a yieldfactor of 0.5; OH$^\bullet$-scavenging rate-coefficient from Fig. 14. **B:** As for A, and: half of the OH$^\bullet$ produced react with HCO_3^- to produce $HCO_3^\bullet/CO_3^{\bullet-}$; $HCO_3^\bullet/CO_3^{\bullet-}$-scavenging rate-coefficient are from Fig. 16

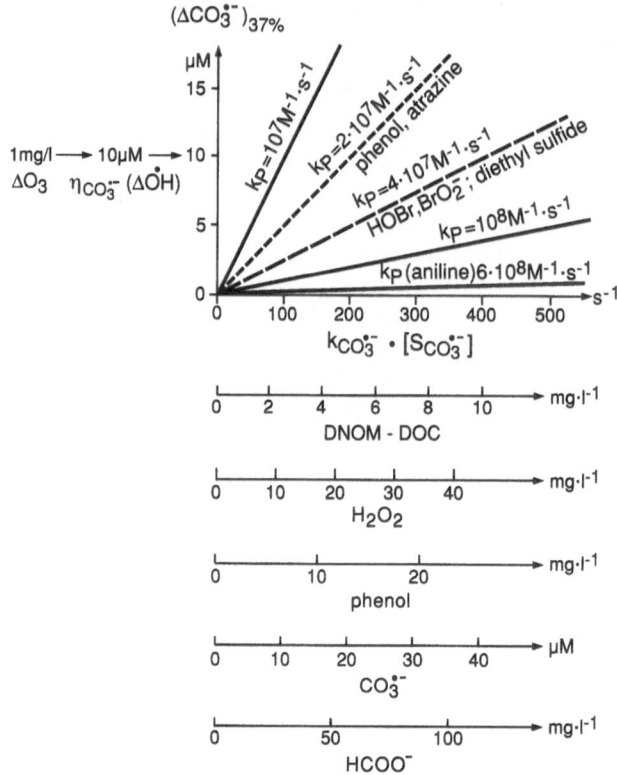

Fig. 16. $(CO_3^{\cdot-})_{37\%}$ required to eliminate pollutants P that react with different reaction-rate constants to 37% of the preceeding concentration. The values are plotted vs the rate constant with which $HCO_3^-/CO_3^{\cdot-}$ become scavenged. The series of abscissa compares the concentration effects of different solutes acting as $HCO_3^-/CO_3^{\cdot-}$ scavengers. The rate constants are from Ref. [14], except for DNOM-DOC and formate ion ($HCOO^-$) that are from [25] and [70], respectively

In Fig. 16, the top abscissa is scaled for the pseudo first-order rate-constant for the reduction of carbonate radicals. Comparative different abscissa are scaled for concentrations of different relevant carbonate radical scavengers that lead to a corresponding scavenging rate. The comparison of the scales allows to conclude that in a common type of freshwater the lifetime of carbonate radicals is dominantly controlled by the reaction with DNOM. E. g., presence of 2 mg/l of DNOM may consume $HCO_3^-/CO_3^{\cdot-}$ with a rate constant of 80 s^{-1}. In natural waters other compounds seem to exhibit a smaller role. We must, however, keep in mind that the rate constant for reduction of carbonate radical by DOM has so far been based only on a first estimate [25].

When we compare this scaling of the abscissa for carbonate radicals given in Fig. 16 with corresponding scales for OH-radicals given in Fig. 14, we can conclude that, in a freshwater, the rate for consumption of carbonate radicals is about 1000 times slower than that for OH$^{\cdot}$. Therefore, in bicarbonate-containing freshwater that contains some DNOM and in which, e.g. 10% of OH$^\cdot$ becomes transformed into the carbonate radical, this may already assume a steady-state concentration that is 100 times that of OH$^\cdot$ (also compare Fig. 15A and B).

The ordinate in Fig. 16 has been scaled for the amount of moles of OH$^\cdot$ per liter to be transferred into carbonate radicals to oxidize a substrate P to a residual concentration of 37% (i.e. to a fraction of e^{-1}). The lines drawn in Fig. 16 represent the corresponding relationship for different oxidizable compounds P and for compounds of different $k_{CO_3, P}$ values. Further compounds of known rate constants [14] can be compared with these examples. These entries show that the high selectivity of carbonate radicals highly limits its importance as an oxidant when comparing it with reactions of OH$^\cdot$. Also when we relate the reactions of this ozonation derived tertiary oxidant with direct reactions of O_3 we can conclude, that the former are of importance only for oxidizing those very exceptional compounds that react readily with carbonate radicals but only slowly with O_3. Hydrogen peroxide, atrazine and hypobromate represent such examples [62, 68]. Exceptions might also occur when AOPs are highly accelerated and lead to an extreme ratio for the concentration of carbonate radical to O_3.

5.5
Role of Further Oxygen-Derived Radicals

During the decomposition of molecular O_3 by the radical-type chain reaction, a sequence of radical-type chain carriers is produced. E.g. some authors conclude from their measurements that during the transformation of O_3 at high pH the concentration of $O_3^{\cdot-}$ might assume a significant steady-state concentration. When $O_3^{\cdot-}$ is protonated, the $HO_3^\cdot$ formed has been deduced to have a lifetime τ of about 10 μs [33, 39]. In many types of waters this lifetime is longer than that of OH$^\cdot$. Hence, the steady-state concentration of $HO_3^\cdot$ becomes even higher than that of OH$^\cdot$ that occurs within the same chain of the succeeding carriers (compare ordinates in Fig. 14). Information on reactivities of $HO_3^\cdot$ is, however, still scarce and so far there are no experimental observations on ozonation processes that require to assume that also $HO_3^\cdot$ acts as a significant oxidant.

Some authors have also assumed $HO_4^\cdot$, that forms within the reaction of OH$^\cdot$ with O_3, to occur as a chain carrier (cf. Scheme 1 and Ref. in [49]). So far, this reactant has also not been considered as an oxidant for solutes for interpreting observations on the oxidations of different products. However, $HO_4^\cdot$ equilibrates with its dissociation products, i.e. with O_3 and OH$^\cdot$ ($\tau(HO_4^\cdot)$ in about 30 μs [32]). Therefore, it could well be that $HO_4^\cdot$ rather acts as a transient reservoir for OH$^\cdot$.

6
Oxidation of Solutes and By-Product Formation

6.1
Dependency on Type of Reactor and Hydraulic Parameters

So far we have described the effect of O_3 and HO$^\cdot$ on the rate of transformation of a compound only for a homogeneous (semi) batch-type or plug-flow reactor cf. Eqs. (11, 19). By denoting the concentration of O_3 or $OH^\bullet$ by [ox] we may summarize for this situation:

$$-\ln [M]_t/[M]_0 = k_M \cdot [ox]_{ss} \cdot t \qquad (24)$$

There, it is the logarithm of the relative residual concentration of a compound M that declines linearly with the concentration of the oxidant, [ox], integrated for the reaction time t. However, in ozonation practice, the hydraulics of the reactor are generally not those of an ideal plug-flow reactor; there is a significant backmixing and short circuiting. For comparison it is therefore useful to integrate the basic rate laws given by Eqs. (10) and (18) also for a completely stirred tank reactor (CSTR). This yields:

$$[M]_t/[M]_0 = 1/(1 + k_M \cdot [ox]_{ss} \cdot t) \sim 1/(k_M \cdot [ox]_{ss} \cdot t) \qquad (25)$$

where the right-hand approximation is for higher elimination factors, where $1 \ll k_M \cdot [ox]_{ss} \cdot t$. Equations (24) and (25) give comparable results only for the beginning of the oxidation process where small elimination factors are considered (about as long as the ratio of $[M]_t/[M]_0$ remains above 1/2). However, if elimination factors should assume a few orders of magnitude, such as required for disinfection processes, the CSTR (see Eq. 25) requires a much higher oxidant exposure value, $[ox]_{ss} \cdot t$ than the batch-type or plug-flow reactor. The rate of disinfection or elimination of a compound M in a batch-type and ideal stirred reactor are compared in Fig. 17.

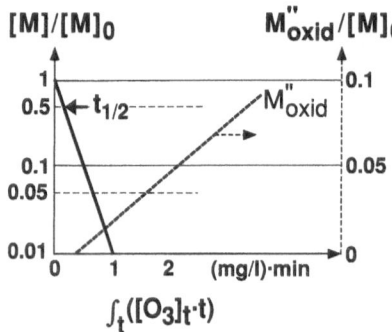

A: batch or plug flow

$$[M]_t/[M]_0 = e^{-k_M[O_3] \cdot t}$$

$$-\ln[M]_t/[M]_0 = k_M \cdot [O_3] \cdot t$$

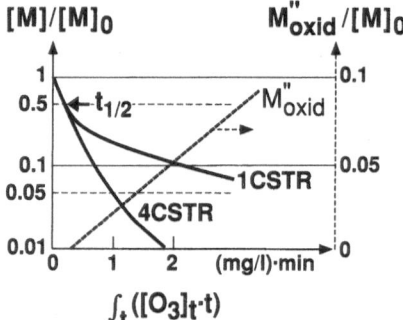

B: continuous stirred tank reactor

$$[M]_t/[M]_0 = \frac{1}{(1+k_M \cdot [O_3] \cdot t)^n}$$

Fig. 17. A formal comparison between the efficiencies of ozonation processes to oxidize a micropollutant M when performed in a batch-type (or plug-flow) reactor and in a completely stirred tank reactor. In the figure on the right *1CSTR* indicates that the reaction proceeds in one single reactor and *4CSTR* indicates a case where the reactor is separated into a series of 4 mixed chambers. n is the number of reactors applied in series. The formation of the by-product M''_{oxid} sketches the appearance of a secondary product (data approximate a case of a bromate formation from bromide)

During extended oxidations, such as required for disinfections in CSTRs, a much higher amount of oxidation by-products can be formed than those achieved in a comparable plug-flow process. Therefore, comparisons of the rate of appearance of an oxidation-by product, M''_{oxid}, here bromate, produced in the presence of bromide (Sect. 6.2), is also included in Fig. 17. The comparison of Fig. 17 A and B show that, also for reasons of by-product formations, well designed reactors for disinfection should rather approach the characteristics of plug-flow reactors without any back-mixing effects. In ozonation engineering practice, the hydraulics of plug-flow reactors have therefore been approached by installing series of stirred reactors. Also inserting vanes, acting as semi-walls, has, since the 1960s allowed many drinking drinking water plants to achieve hydraulics that correspond to series of well mixed chambers. For such a series of n stirred reactors, Eq. (25) becomes:

$$[M]_t/[M]_0 \sim (1/(k_M \cdot [ox]_{ss} \cdot t)^n \tag{26}$$

For $n > 5$ this already approaches the characteristics of a plug-flow reactor (cf. Eq. 24). [71 – 73].

Also the relative factor with which compounds of different reactivity become eliminated within the same oxidation process (of the same $c \cdot$ t-value) changes with the type of the reactor [20]: When Eq. (24), which has been deduced for a (semi) batch-type or plug-flow reactor is applied for calculating the relative depletion of two substrate compounds (M_1 and M_2), it is the ratio of the logarithm of the relative residual concentration of these compounds that becomes, for any $c \cdot$ t:

$$\ln ([M_1]_t/[M_1]_0)/\ln ([M_2]_t/[M_2]_0) = k_{ox,M1}/k_{ox,M2} \tag{27}$$

However, for larger conversion factors achieved in a CSTR, application of Eq. (25) shows that it is now the relative residual concentration (not the logarithm of it) that changes with the ratio of the rate constants:

$$[M_1]_t/[M_1]_0/([M_2]_t/[M_2]_0) \sim k_{ox,M1}/k_{ox,M2} \tag{28}$$

6.2
Formation of Organic By-Products (in the Absence of Bromide)

Ozone and ozonation derived secondary oxidants react with organic and inorganic compounds to produce a large variety of oxidation by-products. Some information on potential product formations from specific precursors may be derived from basic ozone and OH radical chemistry. For this, many reactions of O_3, mostly for non-aqueous solvents, have been reviewed e.g. by Bailey [10, 11] and, emphasizing aqueous reactions, by Larson and Weber [26]. Possible pathways of reactions initiated by $OH^\bullet$ in aqueous solutions have e.g. been reviewed by von Sonntag and Schuchmann [48]. However, such sequences of $OH^\bullet$ reactions were often observed using pulse radiolysis or flash photolysis, where OH· is formed at a high initial concentration. There radical reactions are biased by a dominance of fast radical-radical interactions such as those that dimerize even short lived peroxyradicals to form the tetroxide as a further intermediate. Such intermediates become of minor importance when OH· is produced slowly by transforming ozone.

Complementary, direct observations of product formations that occur during ozonation of isolated representative pollutants have been listed in original studies and compilations [2, 3, 74–78]. There are also publications on products that could be analysed after ozonation (and post-treatment) in many wastewaters and natural waters (see e.g. [3], Chapter 14, [67, 79–86]). Generally, the analytical findings correspond well with the products expected from basic O_3 and OH· reaction chemistry, when this is combined with the knowledge of the fractals of DNOM. However, such observations only serve for empirical exemplifications, because: (i) The main organic reactant for oxidants in freshwaters (DNOM) has a complex and variable structure. (ii) During any ozonation process OH· is formed. This reacts non-selectively and simultaneously with many sites of an organic compound; the large amount of different (and isomeric) organic radicals thereby produced react further, leading to an even larger variety of final reaction products (cf. Fig. 10). (iii) Information on the kinetics that lead to subsequent oxidations of primary and secondary products are not comprehensive. (iv) The processes studied were not sufficiently calibrated in respect to the type and extent of oxidation and in respect to the hydraulics of the reactor. (v) Most aqueous oxidation products are highly polar, and thus there is still a limited analytical capability for analysis. However, despite these difficulties, the formation of the main series of relevant by-products could be characterized by now.

A main problem for discussing product formation during ozonation is that the primary oxidation products become oxidized further or they may decompose (e.g. by hydrolysis). To give an example, the appearance and disappearance of a series of n-alkyl aldehydes during ozonation of lakewater with low ozone dosages is presented in Fig. 18A. In addition, Fig. 18B shows that such product dynamics do not depend solely on the ozone dosage but also on the properties of the water, such as alkalinity. Many further demonstrations for the intermediate build-up of transients have been published [61, 75–80]. Therefore, analytical results can only be generalized when the reactor is specified and when appropriate exposure dose-values, c · t, can be quantified for O_3 and OH· (and sometimes even for HCO_3^- or HOBr, etc.). All the different oxidants that occur during an ozonation process are controlled by different processes and water parameters and they lead to different series of products. For instance, in solutions of concentrated DNOM or decreased carbonate alkalinity, or elevated pH, the transformation of O_3 into OH· is accelerated and the oxidation of substrates becomes shifted in the direction of OH· processes. Even an experienced expert cannot estimate the effect of different oxidants when only the dosage of ozone and the chemical composition of the water are described. However, the fate of specific reference compounds might serve to distinguish between the effects of different oxidants and to calibrate a process in terms of the c · t – values effected by different oxidants.

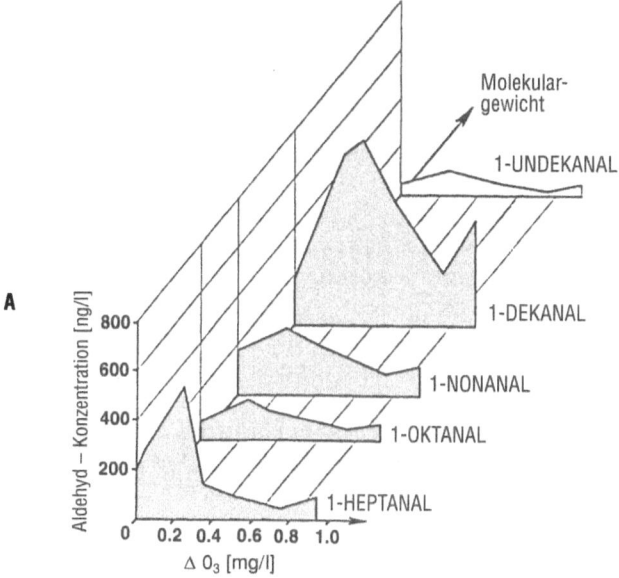

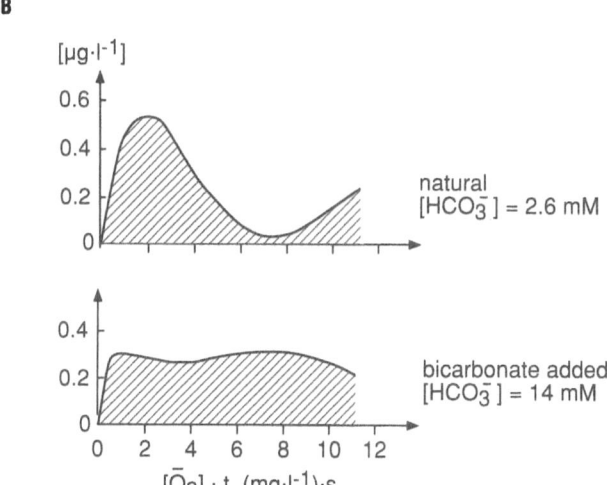

Fig. 18. Dynamics of the formation of some different alkyl aldehydes in lakewater Zurich during ozonation. *Conditions*: DOC = 1.3 mg/l; pH = 7.9. **A:** Concentrations of some specific aldehydes vs the amount of applied ozone. **B:** Concentraion of *n*-decanal as a function of the c_{O_3} · t value in natural water and in similar water that was spiked with additional bicarbonate to scavenge a larger fraction of OH· [67]

The products of ozonation of organic compounds are usually more polar chemicals (e.g. organic acids and carbonyl compounds). Therefore they are more water soluble, less volatile, less lipophilic, less absorbable on activated carbon filters, but generally more degradable by microbiological processes. They also tend to be less odorous and less toxic. Because ozonolysis of olefinic groups, such as present in oleic acids, forms smaller and sometimes more volatile ketones and aldehydes (classical ozonolytic reactions) also new tastes and odors are produced. In natural surface waters containing algal derived products, ozonation may even yield a somewhat acidic (citric) smell, which is only eliminated either through extended ozonation or through a succeeding microbiological treatment step. During an ozonation process even the ratio of molecular weight fractions of DNOM is somewhat shifted to increased fractions of low molecular weight. Due to the formation of carboxylic acid, the acidity of DNOM increases (see e.g. [85]).

Overall, aldehydes, ketones, mixed aldo or keto acids and carboxylic acids are the dominant products that are analyzed (and expected for chemical reasons) [75 – 89]. Epoxides could be important intermediates [6]. In addition, nitro compounds are formed when amines have been oxidized by O_3, but de-amination seems to be a possible reaction when an amino acid is oxidized by OH$^{\cdot}$ (cf. e.g. [61]). Organic sulfides can form sulfoxides or sulfonates.

Ozonation of drinkingwater is generally coupled with a microbiological post-treatment. An assessment of the overall effect of such combined processes on the by-product formation is therefore essential (cf. Sect. 7.4). In this respect, a comprehensive survey of product formations observed in a dozen different types of US water treatment plants and pilot stations gives a useful series of comparable case studies that help for orientation [6, 84]. The waters containing from 1 up to 26 mg/l of TOC had varied concentrations of bromide, and covered a wide pH range. Pollutants were analyzed before and after the ozonation stage, and before and after downstream sand filters and activated carbon colons that included microbiological processes. Also included in the studies were simulations of distribution systems considering post-treatment with chlorine, chloroamine, and chlorine dioxide. However, no information are given on the c · t values for O_3 and OH$^{\cdot}$ and on the hydraulic parameters. The specific classes of analyzed by-products were aldehydes, ketones, carboxylic acids, mixed aldo or keto acids, organic brominated compounds and inorganic bromine species. Also, epoxides and organic peroxides and hydrogen peroxide were identified. The results have shown that conventional dosages of ozone produced different alkyl aldehydes, the sum of which typically assumed a weight fraction of 1 – 2 % of the TOC. About half of this was due to formaldehyde and 20 % to glyoxal and methyl glyoxal. A comparable ozonation would also produce oxalic acid in the order of 1 % per TOC [85]. Hydrogen peroxide and organic peroxides varied over a wide range and assumed up to 1 % of the TOC. All these low-molecular by-products accounted for less than 10 % of all the oxidations that, for stoichiometric reasons, should be expected to be mediated by the dosed ozone. However, this specified series of oxidation by-products is expected to include most compounds that are of health concern (cf. Sect. 6.4). Most of them are well eliminated by subsequent filtrations through activated carbon, slow sand filters, and even by rapid sand filters, provided that these allow for micro-biological processes (also see Sect. 7.4). However, some aldehydes can be reproduced by subsequent addition of disinfectants applied in distribution systems.

6.3
By-Product Formation in Bromide-Containing Water

The best investigated example for a sequence of product formations that proceeds during ozonation and ozone-based AOPs is the oxidation of bromide (Br$^-$) to aqueous bromine (HOBr/OBr$^-$), bromite (BrO$_2^-$), and finally to bromate (BrO$_3^-$) (see Figs. 19, 20).

The product formation in bromide containing waters has for a long time awakened much interest, because it is relevant, (i) when brackish cooling water is treated with ozone, (ii) whenever bromide is added to swimming pool water to produce aqueous bromine or bromamine as a more persistent disinfectant, (iii) in mineral waters that contain elevated concentrations of bromide, (iv) for the ozonation of marine aquaria, (v) when bromide is applied as a (reactive) electrolyte for some electrochemical methods to analyze ozone, (vi) because of its possible interference with some analytical methods that are applied to quantify ozone in water. However, since about 1990, a new priority setting for research on the oxidation chemistry of bromide has mostly been due to the fact that bromate as well as brominated organic compounds have been considered as potential carcinogens.

Br$^-$ concentrations of drinking waters and mineral waters vary from a few micrograms up to several milligrams per liter [91–94]. The major sources of Br$^-$ in freshwaters are related to local geological situations (a rare case) and anthropogenic Br$^-$ emissions from potassium and coal mining or photo proces-

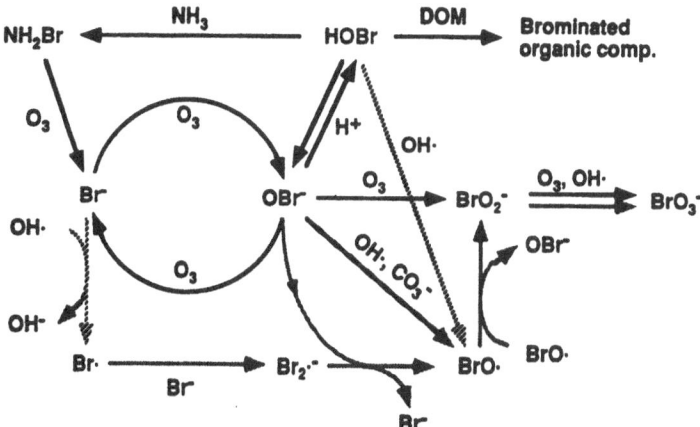

Fig. 19. Reactions upon ozonation of bromide-containing freshwater. Reactions for molecular O$_3$ (*solid lines*) and OH$^{\bullet}$ and HCO$_3^{\bullet}$ radicals (*dashed lines*) are shown [93]

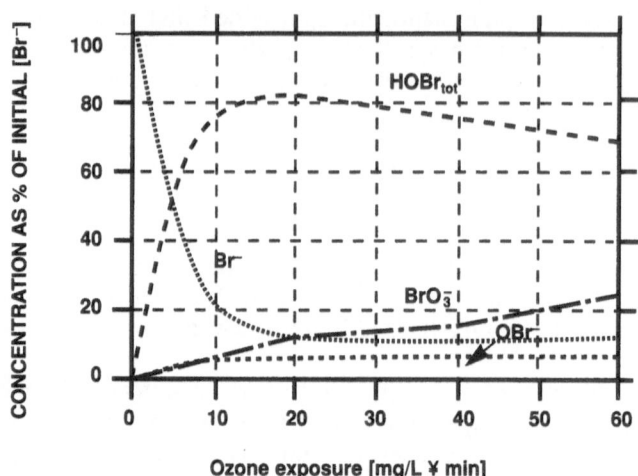

Fig. 20. Model calculations based on 9 reactions for an extended ozonation of a bromide containing water at pH 7 that contains OH˙-scavengers (absence of OH˙ mediated reactions). Concentrations of different species are given in percent of initial [Br⁻] [62]

sing. In coastal groundwaters, elevated Br⁻ concentrations are frequently due to some infiltration of marine water.

Molecular O_3 oxidizes Br⁻ by an oxygen-transfer reaction leading to hypobromite (BrO⁻) (see Eq. 6 and Fig. 19) that quickly assumes the acid-base equilibrium to form hypobromous acid, (HOBr); pK_a (HOBr) = 9. When in the presence of an average O_3 concentration of 0.5 mg/l (typical concentration for drinking water treatment), the time constant τ for the oxidation of Br⁻ is in the range of 1000 s (cf. Fig. 4). As shown in the rather comprehensive model presented in Fig. 19, the aqueous bromine (HOBr/BrO⁻) formed becomes:

- further oxidized, via BrO_2^- to BrO_3^-,
- further oxidized to an intermediate (Br-O-O⁻) that recycles Br⁻,
- reacting with organic solutes to produce brominated organic compounds such as bromoform,
- added to ammonia to form bromamine (NH_2Br),
- reduced by hydrogen peroxide to Br⁻.

In the pH range typical for most freshwaters, a large fraction of the BrO⁻ is protonated to HOBr and thus masked for further reactions with O_3. Therefore, the further oxidation of HOBr/Br O⁻ to BrO_3^- is delayed when the pH is decreased. In the presence of ammonia, a formation of bromamine acts as an additional reservoir for aqueous bromine; bromamine is only slowly oxidized to produce nitrate and to recycle Br⁻ [96]. When there is an absence of an excess of organic compounds, OH˙, or even the more selective carbonate radicals (HCO_3^-/CO_3^{-}) also oxidize aqueous bromine (HOBr/BrO⁻) to BrO˙ that is further transformed to finally produce bromate (Fig. 19). Hence, during AOP enhancing conditions, short lived molecular O_3 sometimes only controls the kinetics of the initial oxidation of Br^- and the final oxidation of BrO_2^-. In this case, the intermediate oxidation steps are rather due to secondary oxidants ($OH^{\cdot}$ or $HCO_3^{\cdot}$). For all these reactions comprehensive sets of rate data are compiled in the literature. References are listed, e.g. in [62]. Model calculations (now based on about 30 reactions) seem to describe well many experimental laboratory results even for the case of AOP-enhancing conditions [93] and when some hydrogen peroxide is present that reduces part of the intermediate HOBr [95]. The basic model has been well tested by extensive and critical experiments (cf. e.g. [62, 97–101]). They have even been successful in interpreting results found on real water treatment plants and can now help to predict bromate formation in different types of waters and when water qualities or ozonation parameters or hydraulic characteristics change [68, 71, 72]. Even an early and restricted set of kinetic data has allowed to interpret the long-range accumulation of bromate in ozonated seawater aquaria of "The Living Seas" [102]. This success in predicting inorganic product formation is due to the fact that, in this case, all relevant inorganic intermediate reactants are well specified and appropriate rate constants have been critically compiled. Series of relevant reactions, such as those due to $OH^{\cdot}$, had thereby already been analyzed within very early studies for describing the effects of ionizing radiation in seawater [14, 62].

In general, the level of the bromate formation during drinking water ozonation is well below the quality criteria of 25 µg/l. A survey that included 38 drinking water plants located in western Europe has shown that bromide concentrations in raw waters ranged from 12 to 200 µg/l [94]. In 11 cases pre-ozonation applied in 12 of these units resulted in less than 2 µg/l bromate. Inter-ozonation, used by 36 of the utilities, produced only in 2 (survey 1993) and 7 plants (survey 1994) more than 5 µg/l bromate. One extreme of 20 µg/l was far above the other values. The authors have concluded that the distributed waters which contain more than 10 µg/l of bromate represent about 20% of the total flow rate of the French waterworks studied. In other treatment plants used for investigating the effect of hydraulics on disinfection parameters and bromate formation, an initial bromide concentration of 50 µg/l did not lead to more than 2 µg/l of bromate when ozone was dosed to attain a $c \cdot t$ value in the range of 0.4 to 6 (mg/l)min [71]. A research group from KIWA (NL), however, concluded that under Dutch ozonation conditions bromate concentrations as high as 12–25 µg/l could be formed, but that an optimization of the processes should allow one to handle a standard as low as 5 µg/l [103]. In a water treatment plant in Halle-Beesen (D) raw water intake must be managed to keep the bromide level between 0.3 and 0.7 mg/l. The authors concluded that, even in this water, and even respecting ozonation disinfection rules, an average of less than 5 µg/l of bromate could be achieved by restricting the ozone dosage and the contact time and working in a low pH range [104]. In Sacramento-San Joan delta river water (TOC: 2 mg/l, pH ~ 8, $[Br^-]$ = 0.1 mg/l) test ozonation with a target ozone residual of 0.35 mg/l resulted in less than 10 µg/l of bromate. But an increase either of bromide or of TOC (that would demand a higher dosage of O_3) would lead to a higher bromate formation [105]. Elevated bromate formations have been reported for ozone based AOP processes, when the ozone dosage was increased to yield a residual amount that allowed one to keep to the disinfection regulations [68]; however, AOP-type processes can only be considered for oxidation processes and not for disinfections. Also, a high bromate yield has to be anticipated for cases where the extent of the ozonation (expressed by the $c \cdot t$ value) is beyond that required for safe disinfection and when the ozonation process is performed in badly designed disinfection plants that show backmixing and therefore have CSTR characteristics (see Sect. 6.1).

In the presence of organic solutes, ozonation of bromide containing waters also produces some brominated organic oxidation by-products. For example, the formation of bromoform, brominated acetic acid, bromoacetonitriles and rather unstable organic bromohydrins (such as 3-bromo-2-methyl-2-butanol) has been quantified for different treatment plants [6, 104–108]. For some of these product formations the sequences of reactions are comparable to those formulated for the reactions that succeed the chlorine induced formation of aqueous bromine in bromide containing waters. The formation of trihalomethanes increases with both, the concentration of TOC and bromide [105, 108]. At extremely elevated ozone doses the production of brominated organic oxidation by-products seems however to decline; a fraction of the aqueous bromine becomes oxidized to the non-reactive bromate and some organic precursors may be oxidized to non-reactive species. Compared with conventional chlorination, conventional ozonation treatment of bromide-containing drinking water produces less brominated haloforms; apparently, ozone also interacts with short lived transients of the haloform formation process and, at extreme ozone dosages, some bromine is already oxidized to bromate.

In the US plants surveyed by the AWWA (cf. Sect. 6.2) [6] the bromoform measured after ozonation was below the 1 µg /l level, even when the bromide concentration reached 0.3 mg/l (non-spiked samples). However, cyanogen bromide was found in a range of up to 20 µg/l (non-spiked samples) and an organic bromohydrin was identified to represent a large proportion of the organic halides formed.

When phenolic compounds are present in bromide-containing water, oxidation of bromide may lead to the formation of bromophenols. But, under drinking water conditions, O_3 degrades phenols and bromophenols much faster than further bromine is supplied from the oxidation of bromide (see entries in Fig. 4 and 6). Therefore, bromophenols do not accumulate during conventional processes. In contrast, when such compounds are formed during chlorination, they persist and make the water, as well as fish and mussels living in it, smell like an unventilated pharmacy.

6.4
Health Effects of Oxidation By-Products

In most raw waters that are of acceptable quality for drinking water production, individual ozonation by-products are formed at such low concentrations that direct tests do not show significant genotoxic, or oncogenic, or other health effects. Therefore, to approach critical concentrations of oxidation by-products, toxicologists sometimes ozonated concentrated extracts of solutes (e.g. of

DNOM) or they used samples or extracts from waters that were treated by extended ozonation. Finding suspicious products in such samples could act as a guide to priority setting for further research. However, on the one hand, O_3 added to solutions of concentrates has a reduced lifetime and the relative role of secondary oxidants becomes increased (cf. Sect. 5). On the other hand, extended ozonation also results in further oxidations of harmful or non-harmful intermediates. Experimental approaches that are based on conditions to enhance by-product formation can therefore only lead to meaningful data when they are complemented by detailed investigations that allow us to quantify the consequences of the changed conditions on the product formation. More straightforward, therefore, are toxicological tests that focus on specified chemical compounds known or expected to be produced during real ozonation, but not safely re-eliminated by the subsequent treatment steps. Results from such studies seem the most valuable; they can also be applied in further fields and be related to other human exposures.

In an instructive chapter, Langlais et al. ([3], Chapter II B) have critically reviewed the concepts of toxicological tests on ozonated water. In addition, evaluations of concepts considering toxicology are continually surveyed at workshops (cf. e.g. [109, 110]). In general we may conclude from such reports that ozonation by-products that are produced in drinking water of acceptable quality appear to be less toxic than those produced by chlorination. Ozonation even transforms many toxic substances, such as many carcinogens and pesticides, to less toxic compounds. (It must however be added, that many types of pesticides are rather difficult to oxidize [29, 63, 111]). Ozonation that corresponded with Dutch drinking water treatment practice did not result in a positive or negative mutagenic activity but comparable chlorination processes had some effects [112]. Complementary experiments on Dutch drinking waters have shown that, dependent on the water quality and treatment conditions, a slight increase of mutagenic activity (Ames tests) occurred, but this was reduced again by increasing the ozone dose.

As summarized above (see Sect. 6.2), in ozonated raw waters that are high in DNOM (2–4 mg/l), compounds such as formaldehyde, glyoxal, methyl glyoxal and many further aldehydes achieve concentrations that are in the 6–30 µg/l range when measured before microbiological post-treatment. But in case of the more widely accepted treatment chains, where ozonation is succeeded by microbiological processes and where post disinfectants are minimalized, the final concentrations of such compounds become decreased (also cf. Sect. 7.4). In addition, comparable series of aldehydes are also found as ubiquitous chemicals in the environment and they are also products of natural metabolic processes: Due to AOPs that also proceed in the atmosphere [12], rainwater contains formaldehyde in the concentration range of 0.1 to 1 mg/l (for a review see Ref. [12, 41]. In vegetables, brewed coffee and many further food and beverages they are found at concentrations that are more than 2 orders of magnitude higher than those reported for ozonated drinking water, even when the latter is measured before microbiological post-treatment [113–115]. For comparison, even for long-term carcinogenicity bioassays, acetaldehyde and formaldehyde had to be administered in drinking water at a concentration that was a few hundred mg/l in order to classify them as potential human carcinogens [116]. Therefore, no significant potential health effect can be attributed to levels of formaldehyde and other carbonyl compounds that, at this time, are analytically found to occur during correct drinking water treatment (also cf. [83]).

Bromate is now considered to be a possible human carcinogen [114, 117]. Its concentration in drinking water giving a human excess lifetime cancer risk of 10^{-5} has been deduced to be 3 µg/l when based on kidney defects observed in feeding experiments with rats. For this, extrapolations have been based on a linearized multistage model, although the genotoxic activity of bromate has been shown only to be indirect and due to a lipid peroxidation, a process which is likely to have a threshold [118, 119]. Based on toxicological linear extrapolations, but accounting for the early crude analytical detection limit for bromate in a natural water matrix, and accepting priority of disinfection safety criteria, the WHO in 1993 recommended a provisional guideline level for bromate in drinking water of 25 µg/l [120] whereas the US-EPA as well as the Drinking Water Commission of the European Union have proposed to set a maximum contaminant level at 10 µg/l [121, 122]. As summarized in Sect. 6.3, for nearly all waters such limits can be respected when optimizing the ozonation process and by observing an adequate raw water management.

7
Combination of Ozone with Other Treatment Processes

In most cases, conventional ozonation and AOP processes are combined with at least one of the downstream treatment processes shown in Fig. 3. For instance, for good drinking water treatment, ozonation is generally followed by a filtration or a microbiological post-treatment to eliminate particulate material and assimilable by-products, respectively. But sometimes, ozonated water is still just post-treated with a further disinfectant (chlorine or chlorine dioxide) to avoid fowling or recontamination in the distribution system. Thus, all ozonation effects and by-products should finally be qualified in combination with the selected post-treatment and considering the water distribution system.

7.1
Effects on Chlorine and Chlorine Dioxide

When O_3 and aqueous free chlorine (HOCl/OCl$^-$) or combined chlorine (e.g. NH_2Cl) are applied in combination, a direct reaction occurs between O_3 and OCl$^-$ or chloramine (NH_2Cl), respectively. In the pH range of drinking water (7–8) and at concentrations of a few mg/l, these oxidants consume each other within minutes (cf. Fig. 4 for rate data). About 77% of the aqueous chlorine reacts with O_3 to finally produce Cl$^-$ and 23% is oxidized to chlorate (ClO$_3^-$) [123–125]. On the one hand, such reactions have already been applied to destroy an O_3 residual. On the other hand, the depletion of aqueous chlorine by an O_3 residual can decrease the efficiency of a post-chlorination.

When chlorine dioxide (ClO$_2$) is applied in water treatment, many organic compounds and reduced inorganic contaminants reduce most of it to chlorite

(ClO_2^-), a suspected blood toxicant. Post-ozonation oxidizes this product quickly to chlorate (ClO_3^-) [28, 123, 125]. Although chlorate now seems to exhibit a lower acute toxicity than chlorite [127], the accepted drinking-water quality-criteria for chlorite and chlorate are comparable. Therefore oxidation of chlorite by ozone cannot be considered any more for upgrading drinking waters that are pre-treated with ClO_2 such as, e.g., to make algae species easier to be withhold by filtration.

7.2
Post-Chlorination: Formation of Halogenated Organic Compounds

O_3 is a very effective oxidant as well for phenolic as for chlorophenolic compounds (see entries in Fig. 6). Therefore, ozonation, before or after chlorination of waters that contain phenol derivatives, leads to a chlorophenol-free water. Historically, this has been the main reason for the early and broad application of ozonation processes in many European water treatment plants that applied phenol-contaminated surface- or groundwaters. Without ozonation chlorine would produce an odor that, in most European regions, is not accepted by the consumers.

Already chemical mechanistic reasoning suggests that ozonolysis of organic compounds might pre-oxidize some further chemical structures that otherwise easily react with chlorine. However, the fractions of O_3 that become transformed into OH· might also produce more hydroxylated organic compounds that exhibit an increased formation potential for chloro compounds. Correspondingly, as often reported, ozonation of natural waters that are high in DNOM has been observed to somewhat lower the formation of halogenated compounds during post-chlorination [3, 5, 6, 127]. For example, pre-ozonation of mesotrophic and eutrophic lakewater decreased the chloroform formation potential to approximately 50%, even when this was not treated by a microbiological process before chlorine addition [128]. However, such limited case-studies do not allow for any generalizations. For example, Baumgart et al. [104] have observed that pre-ozonation of a water of a DOC >2 mg/l and a bromide concentration of 0.3–0.7 mg/l leads to a reduction of the trihalomethanes that were formed within 0.5 hours but not of those that were measured after 24 hours. It must also be noted that pre-ozonation decreases the chlorine demand (although it is not the ammonia that is signifantly oxidized by such a process). This allows for dosing lower amounts of chlorine and this also improves the water quality in respect to the formation of chlorinated compounds. However, pre-ozonation followed by chlorination, particularly when an intermediate microbiological process is missing, leads to the formation of further chlorination by-products: e.g. pre-ozonation has been reported to enhance the chlorine mediated formation of chloralhydrate [127| and in samples from mesotrophic and eutrophic lakes, it increased the formation of trichloronitromethane (chloropicrin) to a few µg/l [128, 129]. This formation increased further when the lakewater was spiked with some amino compounds, and mainly when conditions were changed to enhance OH· radical reactions [128]. Also amino acids become precursors for chloropicrin when ozonation is followed by chlorination [129, 130]. However, chloropicrin is eliminated during subsequent treatment with activated carbon and even by ageing effects. Because the analyzed chloropicrin rather exemplefies potential product formations in general, the given examples may illustrate that the effect of pre-ozonation is often beneficial when haloform is used as a quality parameter, but that it also has a potential of becoming detrimental with regard to further quality criteria.

Much practical experience has been published and reviewed on the reduction of the formation potentials for haloforms and haloacetic acids when the ozonation is combined with a subsequent microbiological process (cf. e.g. [3]), Chapter III.1, [6]). For example, the survey on selected US drinking water units has shown that precursors for trihalomethane formation were effectively removed by ozonation followed by filtrations when these promoted microbiological processes. In contrast, during such microbiological post-treatment precursors for haloacetic acid (HAA) formation were significantly more refractive [6]. From samples of Texas lake waters a comparable water treatment removed up to 50% of the trihaloform precursors and up to 70% of HAA precursors, however, the two lakewaters studied gave different results and even different species of HAAs responded differently [131].

In surface waters that contain some bromide, already a low O_3 dosage appears to oxidize some organic precursors that, during post-chlorination, would otherwise produce halogenated organic compounds. Only a highly extended ozonation, that is not representative for practice, then transforms a relevant fraction of bromide (and hypobromite) to the non-reactive bromate. Upon pre-ozonation post-chlorination then produces less hypobromite that would incorporate into DNOM [see e.g. 106]. In the case of bromide-containing drinking water, precursors for trihalomethane and haloacetic acid formation were better controlled either by ozonation or by biological pre-treatment, respectively. The combined process was effective for the control of all halogenated oxidation by-products [107].

7.3
Effects on Coagulation, Flocculation, Filtration and Flotation

Even in early publications on ozonation effects in lakewaters, the spontaneous flocculation and the creation of larger particles prior to flocculation, the so-called "microflocculation effect", was described [132]. In addition, there has been a steady flux of reports on the effects of pre-ozonation of drinking water and wastewater as related to subsequent coagulation, flocculation, sedimentation, filtration and flotation (ozoflotation) processes. These phenomena have been explained by (i) the increase of the carboxylic acidity of DNOM (or the occurrence of oxalic acid) in presence of calcium ions, (ii), the lysis of algae including the release of cellular polymeric organics, (iii), formation of metastable organics that can polymerize on particle surfaces, (iv) degradation of organic metal complexes with the release of transition metal species (i.e. Fe, Mn), and, (v) degradation of organics absorbed on the particle surfaces (also cf. [133]). Depending on the type of the water, the optimum of such processes can occur at different ozone dosages. So far, direct reactions of molecular O_3 have not been distinguished from reactions of secondary oxidants, and therefore, any ozone-dose relationships cannot be intercompared or quantified. In spite of that, the following examples may illustrate some general experience:

Tobiason et al. [134] reported that pre-ozonation of drinking water lowered the cationic polymer dose but hardly changed the alum dose required for effective turbidity removal by coagulation. Also, ozonation decreased the amount of DNOM that was removed by coagulation but this effect was compensated by an improvement of the filtration through a microbiologically active media. Becker and O'Melia [135] concluded from experiments on model solutions, that the optimum coagulant dose would increase when pre-ozonation transformed a significant fraction of DNOM into carboxylic acids, such as oxalic acid. However, conventional ozonation is known to oxidize only a few percent of the DNOM to oxalic acid [85]. Complementary to this, Chadrakanth and Amy [136] have concluded that an ozone-induced particle destabilization occurred in the presence of calcium that seemed to associate with oxidized constituents of DNOM. This would inhibit adsorption of anionic species onto the alum surface. Ozonation might also produce more ligands on the surface-sorbed DNOM that chelate calcium and thereby reduces the stability of the particles.

Since the mid sixties pre-ozonation has in Switzerland also been applied to improve filtration of algae containing meso-eutrophic lakewater. Thereby different specified algae species respond differently to the process combination [137]. Even low doses of O_3 cause algae to release extracellular organic matter. This may be the reason for an enhanced coagulation. However, at higher O_3 doses this extracellular organic matter is oxidized more completely and coagulation decreases [138].

Pre-ozonation is sometimes applied to improve flotation processes that are used to treat surface waters during algae blooms [137, 138–140] (for a complementary review of earlier lit. cf. [3]).

7.4
Effects on Microbiological Processes

Ozonation transforms fractions of DNOM and many of further organic compounds so that they become better assimilated and degraded by microorganisms. Ozonation of raw waters that contain a significant amount of DNOM may lead to microbial growth and enhanced fouling in water distribution systems (a review is included in a paper by Price [143]). On the other hand, pre-ozonation improves the efficiency of downstream biological water treatment. Therefore, most waterworks in Europe that have applied O_3 for treating surface waters without being equipped with a microbiological post-treatment unit have later retrofitted the system to succeed ozonation by a slow sandfilter, or a so called "biological activated carbon filter" (BAC), or by a reinfiltration into the ground to produce good groundwater. In these systems micro-organisms colonize on the filter media and remove biodegradable or assimilable organic carbon (AOC). Because such combined processes also induce a microbiological nitrification and eliminate ammonia, the purified water also exhibits a lower chlorine demand [144–146]. A review on such combined processes is included in the survey by Langlais et al. [3] and additional publications have been discussed within a paper by Hacker et al. [147] and a comprehensive up-dated series of papers has been included in "Advances in Slow Sand and Alternative Biological Fitration" [148].

DNOM and ozonated DNOM are varying mixtures of diverse compounds of different reactivities (for exemplifications and references cf. [149]). Therefore, ozonation effects on DNOM are not well characterizable. But by experience, ozonation of waters that contain DNOM generally forms aldehydes and keto-aldehydes (see Sect. 6.2 and [149]). Such products have already served well as series of reference compounds to calibrate the microbiological process. Thereby, for instance data for the microbiological degradation of formaldehyde, glyoxal and acetate provided a scale to allocate the removal of readily biodegradable and rather recalcitrant ozonation by-products, respectively [150, 151]. Comparing the microbiological assimilation of series of aldehydes by different systems demonstrated trends similar to those for the removal of assimilable organic carbon (AOC) [150]. A study on the fate of specific aldehydes and keto-aldehydes that were produced during ozonation also allowed one to compare the efficiency of microbiological processes of different US drinking water plants [84]. There the well operated granular activated carbon and sand filters, that supported microorganisms, removed most of aldehydes and ketoaldehydes. In contrast to monoaldehydes, some of the precursors for the formation of halogenated by-products were rather refractive. Subsequent post disinfection using O_3, chlorine or chlorine dioxide has however re-formed significant levels of aldehydes.

An instructive example for the fact that pre-ozonation can either promote or retard biodegradability of contaminants present in an industrial wastewater has been provided by Hu et al. [152] who found that the microbiological elimination of chlorophenol-derived DOC depended primarily on whether or not the activated sludge used in the subsequent biological system had been pre-adapted for eliminating the initial phenolic compounds.

7.5
Effects of Activated Carbon and Carbon Black

In some drinking water plants, residual aqueous O_3 is destroyed by filtration on granular activated carbon (GAC) or by a film of powdered activated carbon (AC), such as sometimes applied as a suspension that floats on top of a slow sand filter to protect the biofilm. Interactions of aqueous O_3 with such black carbon or comparable active surfaces might also occur when traces of carbon fines or carbon black contaminate the ozonation chamber or the ozone impingers. Dusset and Kovacic [153] included, in their publication on the impact of drinking water pre-ozonation on GAC, a rather comprehensive literature review considering the surface chemistry of carbon black. Although, neither the mechanism nor the kinetics of degradation of O_3 that proceed on such surfaces have been well known, the authors have demonstrated by experiments, that even the chemical long-term degradation of GAC was not significantly accelerated when aerated waters passing through the filter also contained an O_3 residual. Their conclusion was that AC degrades O_3 without becoming significantly oxidized. This differs from the experience that aqueous chlorine oxidizes and consumes AC in a nearly stoichiometric process. In addition, we have learnt from extended series of measurements, that the reaction of O_3 on AC or carbon black sufaces initiates exactly the same radical-type chain reaction for transforming O_3 into OH' in the aqueous bulk phase such as elevated pH or as the H_2O_2/O_3 or the UV/O_3 process. For such an AC initiated ozone based AOP we have tentatively proposed the name "Carbozone Process" [154].

8
Conclusions

Well performed ozonation processes result in many water quality improvements. Thereby, our present knowledge allows us to describe adequately most of the chemical reactions that are involved in such processes and to predict oxidation of micropollutants and formation of oxidation by-products.

In freshwater used for drinking water production, the half-life of typical ozone dosages is in the range of 1 to 30 minutes. During conventional drinking water treatment, where a mean concentration of about 0.5 mg/l of ozone is present, the following nucleophilic compounds and chemical structures become oxidized within less than 10 seconds: sulfite, nitrite, olefinic double bonds, phenols, polyaromatic hydrocarbons, organic amines and sulfides. Much slower reactions are found for: bromide, hypobromite, hypochlorite, ammonia, xylene and some partially chloro-substituted olefinic double bonds such as present in some chlorinated solvents. In water treatment practice, compounds such as chloride, benzene, toluene, saturated hydrocarbons, tetrachloroethylene, pyridine and atrazine do not react with molecular O_3.

However, in most types of freshwater radical-type chain reactions transform a large fraction of O_3 into highly reactive secondary oxidants, such as OH'. These transformations proceed with a high stoichiometric yield factor of about 0.5 and they mediate an "Advanced Oxidation Process", (AOP). They are initiated by increased pH, by some dissolved organic material, by activated carbon surfaces, or by hydrogen peroxide. The latter can be dosed directly to the water as a chemical, or it can be produced in situ by UV photolysis of aqueous O_3. The radical-type chain reactions that succeed such intiations are promoted and accelerated by many types of organic compounds that convert non-selective OH' into superoxide anions (O_2^-) or hydrogen peroxide. These react very selectively with further O_3 to produce further OH' radicals and therefore act as chain carriers. However, this chain reaction is quenched by OH' scavenging bicarbonate/carbonate ions and by some types of organic compounds, such as those that contain alkane groups as a main reactant for OH'. The presence of such solutes therefore stabilzes the dosed ozone somewhat. Despite the competition of DOM and bicarbonate (and carbonate ions) for consuming OH', a fraction of OH' is still available for oxidizing even O_3-recalcitrant micropollutants and primary ozonation by-products such as hypobromite, organic acids, aldehydes and ketones. Because of its non-selective and very different types of chemical reactions, occurrence of OH' leads to different sequences of oxidation by-products than molecular O_3. When OH' is scavenged by bicarbonate, also a rather unreactive secondary radical, HCO_3^-, is produced. This may contribute to the oxidation of some exceptional compounds that resist oxidation by O_3, but still yield reactions with HCO_3^-.

Hundreds of reaction-rate constants are now available to characterize the reaction of different types of dissolved compounds with molecular O_3, $OH^{\cdot}$, or further secondary oxidants. Such information has also been acquired from a few other fields of chemistry. Whenever the different types of reactions that occur during ozonation can be distinguished, it becomes possible to apply the wealth of such compilations. This allows us to apply kinetic computer programs even for modelling the overall kinetics of product formation. Such calculations are, however, based on somewhat restricted models that only inlcude the most reprentative reactions. Calculated predictions must, therefore, still be tested and calibrated by well-designed experiments.

In waters that contain dissolved natural organic material (DNOM), ozonation oxidizes DNOM and thereby also produces some organic carboxylic acids, aldehydes, ketones and organic nitro compounds. There, experience shows that the formations of most low-molecular weight compounds are orders of magnitudes below those found in some natural food products and beverages. However, in bromide-containing waters also some hypobromite is formed, which, in the presence of DNOM or ammonia, can yield brominated organics and bromamine, respectively. By further ozonation, it can also be oxidized to bromate. But, except for a few cases, good ozonation practice and adequate water quality management allows waterworks to keep bromate levels below those that are presently set for health reasons.

Many types of impurities enhance the demand of ozone and shorten the lifetime of the dosed ozone and of the secondary oxydants ($OH^{\cdot}$) and increase by-product formation. When setting up an ozonation treatment, it is therefore appropriate to characterize thoroughly the chemical composition of the water and its potential for variations and to provide an appropriate permanent raw water quality protection. Sometimes it is even advisable to pre-treat the raw water. In addition, an ozonation process should generally be complemented by a post-treatment that allows elimination of specific ozonation by-products and microbiologically assimilable compounds that have resulted from oxidized DNOM.

To qualify ozonation by-product formation we must also take into account the effects of all downstream treatment processes, as well as reactions that occur during post-disinfection and within the water distribution system.

Improved analytical methods should allow us to analyze a wider range of by-products and to include more polar compounds. However, further progress in characterizing and comparing disinfection and oxidation processes and by-product formation also requires improved calibration of the ozonation processes with respect to the concentration-time profiles of O_3 and $OH^{\cdot}$ that depend on the chemical composition of the water and the hydraulics of the reactor.

In practice, there is still great potential to improve processes and to describe processes and water qualities in a way that allows useful comparison and generalization of results.

Acknowledgement. I thank Drs. Urs von Gunten and Michael Elovitz for many motivating discussions and suggestions and for correcting the manuscript. My thanks also go to Mrs. Heidi Bolliger for setting up the drawings and to all the staff of EAWAG who has helped and given so much encouragement all along this project, even after my retirement.

References

1. Rice RG, Netzer A, (1982) Handbook of Ozone Technology and Applications. Ann Arbor Science, Ann Arbor. Michigan, USA, p 277
2. Masschelein WJ (1991) Ozone et ozonation des eaux. 2nd edn. Technique et Documentation, Paris. (Engl. edn. (1982): Ozonation Manual for Water and Wastewater Treatment. John Wiley, New York)
3. Langlais B, Reckhow DA, Brink DR(1991) Ozone in Water Treatment, Cooperative Research Report. Amer Water Works Assn and Compagie Générale des Eaux, Lewis, Chelsea Michigan, USA
4. Maier D, Gilbert E, Kurzmann G (1993) Wasserozonung in der Praxis. Oldenbourg, München
5. Singer PhC (1990) J Amer Water Works Assn 82:10
6. Glaze WH, Weinberg HS (1993) Identification and occurrence of ozonation by-products in drinking water. AWWA Res Found, 6666 West Quincy Ave. Denver, CO 80255, USA (ISBN 0-89867-691-6)
7. International Ozone Assn. (1995) Proceedings of the 12th World Congress, May 1995, Lille, France, IOA, c/o Wasserversorgung Zurich, 8023 Switzerland
8. Ozone Science and Engineering. (1997) Rice RG (ed) J of the Intern Ozone Assn Vol 19, Lewis, Michigan, USA
9. Ozone News (1997) Rice RG (ed) Intern Ozone Assn, 1331 Patuxent Drive, Ashton, MD (USA) 20861 (ISBN 1065-5905)
10. Bailey PS (1978) Ozonation in Organic Chemistry. 1: Olefinic Compounds. Academic Press, New York
11. Bailey PS (1982) Ozonation in Organic Chemistry 2: Nonolefinic Compounds. Academic Press, New York
12. Hoigné J (1995) Tribune de l'eau 575:3:31
13. Neta P, Huie E, Ross AB (1988) J Phys Chem Ref Data 17:1027
14. Ross AB, Bielsky BHJ, Buxton GV, Cabelli DE, Greenstock, CL, Helman WP, Huie RE, Neta P (1992) NIST Standard Referenece Database, NDRL/NIST, NIST Standard Ref Data, Gaithersburg, MD 20899 USA
15. Hoigné J (1998) (to be publ.)
16. Wanner HU, Gilgen A (1966) Archiv Hyg. und Bakteriologie 150:78
17. Morgan KT (1995) In: Disinfection by-products in drinking water: Critical issues in health effects research, Workshop Report, Intern. Life Sciences Inst. Washington D.C. 20036-4804, USA
18. Fuhrer J, Achermann B (1994) UN-ECE Workshop Report, Schriftenreihe Swiss Fed. Res. Station for Agricultural Chemistry and Environmental Hygiene, CH-3097, Liebefeld-Bern
19. Bard AJ, Parsons R, Jordan J (1985) Standard Potentials in Aqueous Solution. (IUPAC) Marcel Dekker, New York
20. Hoigné J (1982) In: Rice RG, Netzer A (ed) Handbook of ozone technology and application, I: 341, Ann Arbor Science, Ann Arbor. Michigan, USA
21. Horn RJ, Straughton JB, Dyer-Smith P, Lewis DR (1996) Ozone Sci Engin 18:57
22. Hoigné J (1988) In: Stucki S (ed) The chemistry of ozone in process technolgies for water treatment. Plenum, New York
23. Tomiyasu H, Fukutomi H, Gordon G (1985) Inorg Chem 24:2962
24. Logager T, Holcman J, Sehested K, Pedersen T (1992) Inorg Chem 31:3523
25. Larson RA, Zepp RG (1988) Environ Toxic Chem 7:265

26. Larson RA, Weber EJ (1994) Reaction mechanism in organic chemistry. Lewis, Chelsea, Michigan, USA, p 313
27. Hoigné J, Bader H (1983) Water Res 17:173 and 185
28. Hoigné J, Bader H, Haag WR, and Staehelin J (1985) Water Res 19:993
29. Yao CCD, Haag WR (1991) Water Res 25:761
30. Andreozzi R, Insola A, Caprio V, D'Amore MG (1991) Water Res 25:655
31. Staehelin J, Hoigné J (1982) Environ Sci Technol 16:676
32. von Piechowski AMJ (1991) Der Einfluss von Kupferionen auf die Redoxchemie des atmosphärischen Wassers. Thesis, Eidgen. Techn Hochschule Zürich (ETH – Z) (Dissertation ETH Nr. 9512)
33. Bühler R, Staehelin J, Hoigné J (1984) J Phys Chem 88:2560
34. Staehelin J, Bühler RE, Hoigné J (1984) J Phys Chem 88:5999
35. Bahnemann D, Hart EJ (1982) J Phys Chem 86:252
36. Holcman J, Sehested K, Bjergbakke E, Hart EJ (1982) J Phys Chem 86:2069
37. Sehested K, Holcman J, Bjergbakke E, Hart EJ (1984) J Phys Chem 88:4144; (1982) ibid. 86:2066
38. Sehested K, Holcman J, Bjergbakke E, Hart EJ (1987) J Phys Chem 91:2359
39. Nauser TN (1996) Einfluss von Kupferionen und organischen Puffern auf den Ozonabbau in Wasser; Kinetische Untersuchungen. Thesis, Eidgen. Techn Hochschule Zürich (ETH – Z) (Dissertation ETH Nr. 11461)
40. Stumm W, Morgan JJ (1996) Aquatic chemistry; chemical equilibria and rates in natural waters, 3rd edn. Wiley-Interscience, New York
41. Jans U (1996) Radikalbildung aus Ozon in atmosphärischen Wassern; Einfluss von Licht, gelöster Stoffe und Russpartikel. Thesis, Eidgen. Techn Hochschule Zürich (ETH – Z) (Dissertation ETH Nr. 11814); Jans U, Hoigné J (1998) Ozone Sci Engin, in press
42. Reckhow DA, Knocke WR, Kearns MJ, Parks CA (1991) Ozone Sci Eng 13:675; Toui Syuji (1991) Ozone Sci Engin 13:623
43. Gurol MD, Neloulnalni S (1984) Industr Engin Chem Fundamentals 23:54
44. Perez RR, Gomez MM and Ramos LR, Ozone reactions with carbohydates in aqueous medium in Proc. 8th Ozonc World Congress, SepL 1987, Zürich. Internalional Ozone Associalion (IOA). (Unionsverlag, Zürich) 1987, Vol. 2, pp E106–27
45. Hoigné J, Bader H (1979) Ozone Sci Engin 1:357
46. Hoigné J, Bader H (1994) Ozone Sci Engin 16:121
47. Sehested K, Holcman J, Bjergbakke E, Hart EJ (1991) Environ Sci Technol 25:1589; Sehested K, Corfitzen J, Holcman J, Fischer, CH Hart EJ (1982) J Phys Chem 86:2066
48. von Sonntag C, Schuchmann, H-P (1991) Angew Chem Int Ed Engl 30:1229 (German version: Angew Chem 103:1255); (1998) Peroxyl Radicals in Aqueous Solution. In: Alfassi ZB (Ed.)
49. Staehelin J, Hoigné J (1985) Environ Sci Technol 19:1206
50. Taube H (1956) Trans Farad Soc 53:656; Peroxyl Radicals, John Wiley Ltd. (in press)
51. Chelkowska K, Grasso D, Fabian I, Gordon G (1992) Ozone Sci Eng 14:33
52. Peyton GR, Glaze WH (1987) In: Zika RG, Cooper (eds) ACS Symp Ser 327:76, ACS, Washington DC)
53. Hoigné J (1997) Water Sci Techn 36 35(4):1–8
54. Glaze WH, Peyton GR, Lin S, Huang FY, Burleson JL (1982) Environ Sci Techn 16:454
55. Peyton GR (1990) In: Ram NM, Christman RF, Cantor KP (eds) Significance and treatment of volatile organic compounds in water supplies. Lewis, Chelsea, Michigan, USA, p 339
56. Glaze WH (1987) Environ Sci Technol 21:224
57. Hoigné J, Bader H (1987) In: Proc. 8th Ozone World Congress, 15–18 Sept., Zurich, Intern Ozone Assoc., c/o Wasserversorgung Zurich (ISBN 3-293-00126) p K 83
58. Peyton GR, Glaze WH (1988) Environ Sci Technol 22:761
59. Masten SJ, Hoigné J (1992) Ozone Sci Eng 14:197
60. De Laat J, Tace E, Doré M (1994) Water Res 28:2507
61. Le Lacheur R, Glaze WH (1996) Environ Sci Technol 30:1072
62. von Gunten U, Hoigné J.(1994) Environ Sci Technol 28:1234

63. Haag WR, Yao CCD (1992) Environ Sci Technol 26:1005
64. Hoigné J, Bader H (1977) Vom Wasser 48:283
65. Nowell LH, Hoigné J (1992) Wat Res 26:599
66. Haag WR and Yao CCD (1993) In: Proceedings 11th Ozone World Congress, San Francisco (Intern Ozone Assn)
67. Zürcher F, Bader H, Hoigné J (1982) Concerted action analysis of organic micropollutanls in water (Cost Project 64 bis), 2:198 Commission of the European Communities, Brussels
68. von Gunten U, Bruchet A, Costentin E (1996) J Amer Water Works Assn 88:53
69. Glaze WH, Beltran F, Thukanene T, Kang J-W /1992) Water Poll Res J Canada 27:23
70. Zellner R, Herrmann H, Exner M, Jacobi H-W, Raabe G, Reese A (1996) In: Warneck P (ed) EUROTRAC Vol 2, Heterogeneous and liquid phase processes. Springer, Berlin Heidelberg New York
71. Roustan M, Duguet JP, von Gunten U, Lainé JM, Mallevialle J (1995) Bromate formation impact of ozone contactor hydraulics and operating conditions; 12th World Congress, Intern. Ozone Assoc., Lille France (Intern. Ozone Assoc., c/o Wasserversorgung Zürich, CH), p 201
72. Roustan M, Duguet JP, Lainé JM, Do-Quang, Mallevialle J (1996) Ozone Sci Techn 18:87
73. Henry, DJ, Freeman EM (1995) Ozone Sci Engin 17:587
74. Mallevialle J, Duget JP (1991) In: Masschelein WJ (ed) Ozone et ozonation des eaux, Technique et Documentation, Paris, p 121
75. Laplanche A, Martin G (1991) In: Masschelein WJ (ed) Ozone et ozonation des eaux, Technique et Documentation, Paris
76. Gilbert E (1983) Ozone Sci Engin 5:137
77. Gilbert E (1992) Water Res 26:1533
78. Gilbert E, Hodenberg S, Klinger J, Eberle SH (1995) in: Proc. 12 World Congress, Intern. Ozone Assoc., Lille, France (Intern. Ozone Assoc., c/o Wasserversorgung Zuerich, CH)
79. Brunet R, Bourbigot MM, Doré M (1984) Ozone Sci Engin 6:163
80. Huck PM, Anderson WB, Rowley SM, Daignault SA (1990) J Water SRT – Aqua 39:321
81. Krasner StW, Glaze WH, Weinberg HS, Daniel PhA, Najm IN (1993) J Amer Water Works Assn 85:1:73
82. Krasner StW, Sclimenti MJ, Bradley MC (1993) J Amer Water Works Assn 85(5):62
83. Glaze WH, Koga M, Cancilla D (1989 b) Environ Sci Technol 23:838–847
84. Weinberg HS, Glaze WH, Krasner StW, Sclimenti MJ (1993) J Amer Water Works Assn 85(5):72
85. Bose P, Bezbarua BK, Reckhow DA (1994) Ozone Sci Engin 16:89
86. Andrews SA, Huck PM (1994) Ozone Sci Engin 16:1
87. Najm SN, Krasner StW (1995) J Amer Water Works Assn 87(1):106
88. Glaze WH, Koga M, Cancilla D, Wang K, McGuire MJ, Liang Sun, Davis MK, Tate CH, Aieta EM (1989 a) J Amer Water Works Assn 81(8):66
89. Garcia-Araya JF, Croué JP, Betran FJ, Legube B (1995) Ozone Sci Engin 17:647
90. Bull RJ, Koepler FC (1991) In: Health effects of disinfectants and disinfection by-products. AWWA Res. Found., American Water Works Association (ISBN 0-8986-7-566-9)
91. Haag WR, Hoigné J, Bader H (1982) Vom Wasser 59:237; Haag WR, Hoigné J (1983a) Environ Sci Techn 17:261
92. von Gunten U, Hoigné J (1992) J Water SRT – Aqua 41:299
93. von Gunten U, Hoigné J (1996) In: Minear R, Amy GL (eds) Disinfection by-products in water treatment, CRC Press, Boca Raton, p 187
94. Legube B (1996) Ozone Sci Eng 18:325; Legube B, Bourbigot MM, Deguin A, Kruithof JC, Fielding M, Mallevialle J, Malia L, Mordiel A, WilbournJ (1995) Proc. 12th World Congress, Intern. Ozone Assoc., Lille France (Intern. Ozone Assoc., c/o Wasserversorgung Zürich, CH), p 129
95. von Gunten U, Oliveras Y (1997) Water Res 31:900
96. Haag WR, Hoigné J, Bader H (1984) Water Res 18:1125
97. Song R, Minear R, Westerhoff P, Amy G (1996) Environ Techn 17:861

98. Krasner STW, Glaze WH, Weinberg HS, Daniel PhA, Najm IN (1993) J Amer Water Works Assn 85(1):73
99. Siddiqui MS, Amy GL, Rice RG (1995) J Amer Water Works Assn 87(10):58
100. Westerhoff P, Amy G, Song R, Minear R (1996) In: Disinfection By-Products in Water Treatment, Minear R, Amy GL (eds) CRC Press, Boca Raton, p 255
101. Croué JP, Koudjonou BK, Legube B (1996) Ozone Sci Engin 18:1
102. Grguric G, Trefry JH, Keaffaber JJ (1994) Water Res 28:1087
103. Kruithof JC, Noordsij A, Puijker L.M, van der Gaag MA (1985) In: Jolley RL (ed) Water chlorination, environmental impact and health effects, 5:1137, Lewis, Chelsea, Michigan USA; Kruithof JC, Oderwald-Müller EJ, Mejers RT (1995) In: Proc. 12th World Congress, Intern. Ozone Assoc., Lille, France (Intern. Ozone Assoc., c/o Wasserversorgung Zuerich, CH)
104. Baumgart W, Weber A, Schmidt W, Wricke B, Kühn W (1995) Proc. 12th World Congress, Intern. Ozone Assoc., Lille France (Intern. Ozone Assoc., c/o Wasserversorgung Zürich, CH), p 251
105. Krasner StW, Sclimenti MJ, Means EG (1994) J Amer Water Works Assn 86(6):34
106. Shukary HM, Miltner RJ, Summers RS (1995) J Amer Water Works Assn 86(6):72
107. Shukairy HM, Miltner RJ, Summers RS (1994) J Amer Water Works Assn 87(10):71
108. Siddiqui M, Amy G (1993) J Amer Water Works Assn 85:1:63
109. Bull RJ, Koepler FC (1991) In: Health effects of disinfectants and disinfection by-products. AWWA Res Found, American Water Works Association (ISBN 0-8986-7-566-9)
110. Intern. Life Sciences Inst. (1995) Disinfection by-products in drinking water: Critical Issues in Health Effects Research, Workshop Report, Washington D.C. 20036–4804, USA
111. Meijers RT, Oderwald-Muller EJ, Nuhn PANM, Kruithof JC (1995) Ozone Sci Engin 17:673
112. Kool HJ, Hrubec J, van Kreijl CF, Piet GJ (1985) The Sci of the Total Environ 47:229
113. World Health Organisation (1989) Environmental Health Criteria 89, Formaldehyde. WHO, Geneva CH; World Health Organisation (1991) IARC Monographs on the Evaluation of Carcinogenic Risk to Humans, WHO Intern. Agency for Research and Cancer, Vol 51
114. World Health Organisation (1990) IARC Monographs on the Evaluation of Carcinogenic Risk to Humans, WHO Intern. Agency for Research and Cancer, Vol 52
115. Ames BN, Magaw R, Gold LS (1987) Science 236:271
116. Soffritti M, Belpoggi F, Ciliberti A, Bortoluzz L, Santi A, Lenzi A (1995) In: Disinfection by-products in drinking water: Critical Issues in Health Effects Research, Workshop Report, Intern. Life Sciences Inst. Washington D.C. 20036–4804, USA
117. Kurukawa Y, Mekawa J, Takaashi M, Kokube T (1986) J Natl Inst 77:977
118. O'Neill G, Nunn J, Bacelo S and Chipman JK Fawell J, (1995) In: Disinfection by-products in drinking water: Critical Issues in Health Effects Research, Workshop Report, Intern. Life Sciences Inst. Washington D.C. 20036–4804, USA
119. Fawell J, O'Neill G (1995) In: Disinfection by-products in drinking water: Critical Issues in Health Effects Research, Workshop Report, Intern. Life Sciences Inst. Washington D.C. 20036–4804, USA, p 50
120. World Health Organisation (1993) Guidelines for drinking water quality, WHO Geneva, CH
121. US Fed. Reg. (1994) 5(145):38668
122. Amtsblatt der Europäischen Gemeinschaften (1995) Nr. C 131/5
123. Richard Y, Brener Y (1982) In: Rice RG, Netzer A (ed) Handbook of ozone technology and application, I: 277 Ann Arbor Science, Ann Arbor, Michigan, USA
124. Haag WR, Hoigné J (1983 b) Water Res 17:1397
125. Hoigné J (1985) Gas-Wasser-Abwasser 65:773
126. Romano R (1995) In: Disinfection by-products in drinking water: Critical Issues in Health Effects Research, Workshop Report, Intern. Life Sciences Inst. Washington D.C. 20036–4804, USA p 53

127. Singer Ph, Harrington GW, Gretchen A, Cowmen A, Maria E (1995) Proc. 12th World Congress, Intern. Ozone Assoc., Lille France (Intern. Ozone Assoc., c/o Wasserversorgung Zürich, CH), p 115
128. Hoigné J, Bader H (1988) Water Res 22:313
129. Becke Ch, Maier D, Sontheimer H (1984) Vom Wasser 62:125
130. Duguet JP, Broadart E, Duissert, B, Mallevialle J (1985) Ozone Sci Engin 7:241
131. Speitel GE Jr, Symons JM, Diehl AC, Sorensen HW, Cipparone LA (1993) J Amer Water Works Assn 85(5):86
132. Maier D (1979) In: Oxidation technique in drinking water treatment, EPA-570/9-79-020, p 394; (1984) In: Rice RG, Netzer A (eds) Handbook of Ozone Technology and Application, Vol. II, Butterworth, p 123
133. Jekel MR (1994) Ozone Sci Engin 16:55
134. Tobiason JE, Reckhow DA, Edzwald JK (1995) J Water SRT – Aqua 44:142
135. Becker WC, O'Melia ChR (1996) Ozone Sci Engin 18:311
136. Chandrakanth MS, Amy GL (1996) Environ Sci Technol 30:431
137. Aeppli J, Gros H (1991) In: Masschelein WJ (ed) Ozone et ozonation des eaux, Technique et Documentation, Paris [138. Richard Y (1993) Ozone Sci Engin 15:465
138. Paralkar A, Edzwald JK (1996) J Amer Water Works Assn 88(4):143
139. Richard Y (1993) Ozone Sci Eng 15:465
140. Wilson D, Lewis J, Nogueria F, Faivre M, Boisdo V. (1993) Ozone Sci Engin 15:481
141. Bourbigot MM, Martin N, Faivre M, Le Corre K, Quennell S (1991) J Water SRT – Aqua 40:86
142. Faivre M, Langlais B (1991) in: Masschelein WJ (ed) Ozone et ozonation des eaux, Technique et Documentation, Paris
143. Price ML, Bailey RW, Enos MH, Hermanowicz SW (1993) Ozone Sci Engin 15:95
144. Schalekamp M (1977) Gas-Wasser-Abwasser 57:657
145. Sontheimer H, (1978) J Amer Water Works Assn 70(7):393
146. Wang BZ, Tian JZ, Yin J (1986) Aqua 6:351
147. Hacker PA, Paszko-Kolva Chr, Stewart MH, Wolfe R. L, Means EG (1994), Ozone Sci Engin 16:197
148. Nigel G, Collins R (1996) Advances in slow sand and alternative biological filtration. John Wiley, pp 461; Abstracts: Ozone News (1996) 24(3):15
149. Schechter DS, Singer PhC (1995) Ozone Sci Engin 17:53
150. Krasner St W, Sclimenti MJ, Coffey BM (1993) J Amer Water Works Assn 85(5):62
151. Huck PM, Anderson WB, Rowley SA, Daignaul H (1990) J Water SRT – Aqua 39:321
152. Hu Szu-Tsong, Yu Yue-Hwa (1994) Ozone Sci Engin 16:13
153. Dussert W, Kovacic SL (1997) Ozone Sci Engin 19:1
154. Jans U, Hoigné J (1998) Ozone Sci Engin 20 (in press)

Transformation of Aqueous Pollutants by Chlorine Dioxide: Reactions, Mechanisms and Products

Chaim Rav-Acha

State of Israel, Ministry of Health, Research Laboratory of Water Quality, P.O.Box 8255, Abu-Kabir Tel-Aviv 61082, Israel
E-mail: vglezer@vms.huji.ac.il

Dedicated to my admired teacher and friend, Professor Shalom Sarel

Chlorine dioxide is frequently used for the disinfection of drinking water and wastewater effluents. Its use is desired in particular in case of taste and odor problems caused by chlorine, or when the concentration of undesirable chlorinated by-products exceeds their maximum allowable concentrations. The chemical reactions of chlorine dioxide are much more specific and selective than those of chlorine and, as a consequence, it has a lower demand by water and does not produce as many by-products. However, chlorine dioxide is not an ideal oxidant either, because it reacts with various inorganic and organic aquatic compounds to produce a variety of products. The present chapter summarizes the reactions of ClO_2 with the major inorganic compounds or organic groups which are, or may be, present in water, and describes their products. Special attention has been paid to the mechanisms of reactions in order to enable the prediction of reactions and products not specified in the current chapter. Some toxicological assessments of chlorine dioxide disinfection in general, and of some of its by-products in particular, are also included.

Keywords: chlorine dioxide: – properties; – reactions with phenols, amines, alkenes, PAH; Disinfection by-products (DBPs); chlorite, chlorate.

Contents

The Handbook of Environmental Chemistry Vol. 5 Part C
Quality and Treatment of Drinking Water II (ed. by J. Hrubec)
© Springer-Verlag Berlin Heidelberg 1998

List of Symbols and Abbreviations

E.T.	Electron Transfer
GAC	Granular Activated Carbon
MAC	Maximum Allowable Concentration
MCL	Maximum Contamination Level
MO	Molecular Orbital
PAH	Polynuclear Aromatic Hydrocarbons
THM	Trihalomethanes
TLV	Threshold Limit Value

1
Introduction

It is now known that chlorine, which was the traditional disinfectant for drinking water over decades, has some serious disadvantages, the most significant of which is the formation of halogenated organic compounds. Some of these compounds are suspected carcinogens and mutagens, and some may have other undesirable health effects [1, 2, 3].

The formation of halogenated organics is particularly significant under circumstances, like these prevailing in Israel, where high chlorine concentrations are required. Such circumstances may prevail when it is necessary to maintain residual disinfectant in long distribution lines with many extensions, or when the pH of water is relatively high (> 8) and therefore the hypochlorite ion, which is a less efficient disinfectant, predominates over hypochlorous acid. High levels of halogenated organics are also generated if the water contains bromide in high concentration. In such case, chlorine oxidizes bromide to bromine, which, in turn, produces more halogenated (brominated) compounds than does chlorine [68]. Under all the above-mentioned circumstances, one may consider the use of alternative disinfectants, among which chlorine dioxide is one of the most promising.

Apart from producing less halogenated compounds in relation to chlorine, advantages of chlorine dioxide include:

a) surface waters typically have a lower demand for ClO_2 than for Cl_2, so a higher residual of disinfectant can be preserved [4];

b) unlike chlorine, it does not produce (or produces very few) chlorophenols that cause taste and odor problems [48];

c) it is a more efficient disinfectant, especially against viruses [5, 6];

d) the equipment required for ClO_2 use is convenient to install, operate and maintain [7] – it is also safer to use, since it does not require large reservoirs, as does chlorine, thereby eliminating the danger of an explosion occurring near a populated vicinity;

e) under regular water treatment conditions, it does not hydrolyze, and therefore its oxidation potential and disinfection strength are not pH dependant
like those of chlorine – consequently it may be superior to chlorine, particularly for waters of high pH.

In spite of the above-mentioned advantages, the use of ClO_2 for water disinfection is not altogether favorable, due to the fact that ClO_2 also produces by-products that may have some undesirable health effects. The main inorganic products of ClO_2 during water disinfection are chlorite, chlorate and hypochlorous
acid. The main health concern is that chlorite and chlorate may cause hemolytic anemia in sensitive people [8, 9].

Very little is known about organic by-products when drinking water are
disinfected with chlorine dioxide. Some information is available from the in-
house investigation of Stevens on Ohio River water [10]. Stevens compared the
total ion current profiles obtained by GC/MS of the purgable compounds formed in ClO_2-treated and untreated water and found an excess of low molecular
weight $C_1 – C_9$ aldehydes in the treated water. De Greef and coworkers [91] also
found some aldehydes together with ketones, phenols and a benzoquinone
derivative, while studying the effect of ClO_2 disinfection on river Rhine waters
in the Netherlands. Carboxylic acids, which were expected to be produced by
further oxidation of the aldehydes by ClO_2, were not detected in those studies,
probably because only the volatile compounds could be detected by the techniques used without derivation.

However, recently carboxylic acids have been identified as ClO_2 by-products
when extraction and on-column concentration techniques were used, rather
than the purge and trap method, to identify these by-products by Richardson et
al. [92]. Among the unusual by-products in that study were a series of maleic
anhydride derivatives, the source of which is unclear. One possible explanation
could be their production as phenol oxidation products.

2
Physical Properties, Molecular Structure and Chemical Affinity of Chlorine Dioxide

At ordinary temperature, chlorine dioxide is a gas of yellow-orange color, with
a sharp chlorine-like odor. Its freezing point is $-59\,^\circ C$, and its boiling point is
$9.7\,^\circ C$ [11]. In liquid form, it has a deep red color [12]. Chlorine dioxide gas is a
strong irritating agent with a TLV (threshold limit value for industrial inhalation) of 0.1 ppm, as opposed to 1 ppm for chlorine. Its explosion level is 4% in
air, above which it is detonated by sparks [11].

At $25\,^\circ C$ chlorine dioxide is about 23 times as concentrated in the aqueous solution as in the gas phase with which it is in equilibrium [13, 14]. The partition
coefficient of ClO_2 between water and the gaseous state is 70 ± 0.7 at $0\,^\circ C$, 45 at
$15\,^\circ C$ and 21.5 ± 0.76 at $35\,^\circ C$ [15]. Chlorine dioxide is, thus, about ten times
more soluble in water than chlorine.

Aqueous solutions of ClO_2 are quite stable when kept in a dark bottle at $4\,^\circ C$.
However, in a distribution network, it is not possible to maintain a high residual

of ClO_2 over a long period of time, because it decomposes slowly into chlorine and oxygen [16]. Another disproportionation of ClO_2, into chlorate plus chloride under light radiation, is also possible [17]. The latter predominates in alkaline medium.

At neutral pH, chlorine dioxide does not tend to hydrolyze. However, a few polyhydrates of ClO_2 have been reported for concentrated aqueous solutions [18, 19]. The hydrates have a limited solubility and revert to the unhydrated form above 18 °C. Solid chlorine dioxide polyhydrate can be handled safely at low temperatures, usually in the form of blocks encased in ice coatings, which are frequently shipped cold and regenerated as a gas when needed [15].

The chlorine dioxide molecule has a C_{2V} symmetry and contains two ClO links, which predominantly show double-bond character and have a ClO distance of 1.47 Å and a OClO angle of $117 \pm 1.7°$ [15, 20, 21]. These parameters explain the dipole moment of ClO_2, which is 1.784 ± 0.01 D [22].

The U.V. absorption spectrum of ClO_2 possesses a broad absorption band, with a maximum near 3600 Å and a molar extinction coefficient of about 1250 M^{-1} cm^{-1}, which is practically not affected by pH, temperature or ionic strength, and is therefore often used for the quantitative determination of ClO_2.

Conforming with its molecular structure is the infra-red spectrum of ClO_2 which, in its ground state, possesses the fundamental frequencies: $v_1 = 945$ cm^{-1}; $n_2 = 445$ cm^{-1} and $n_3 = 1105$ cm^{-1}. The bending vibration at 445 cm^{-1} exhibits a doublet character separated by $\Delta V = 3.5$ cm^{-1} and is accompanied by rotational lines which are resolved in the R-side of the band. Many additional bands in the infra-red spectrum have been observed and discussed by Nielsen and Woltz [23].

Chlorine dioxide contains 19 valence electrons and therefore possesses paramagnetism with a molar susceptibility of 1310×10^{-6} cgs units per mol. Nevertheless, unlike other free radicals with an odd number of electrons, such as OF or CN, which are unstable and therefore tend either to decompose or to dimerize, chlorine dioxide is quite stable. It does not tend to decompose (or decomposes very slowly) and does not dimerize at all. This was rationalized by Green and Linnett (24), based on two types of approach. The first approach, based on MO theory, places the unpaired electron of ClO_2 in a molecular orbital having a b_1 symmetry. Such a molecular orbital has the plane of the molecule as a node, and therefore no σ bond can be formed by pairing two electrons in the b_1 orbital. Therefore dimerization is avoided.

According to the second approach, based on valence bond theory, a dimerization can occur only when the number of bonds in the dimer exceeds those of both monomers. Thus, for example, when two CN radicals dimerize to cyanogen:

$$2 \cdot C\equiv N| \ (\text{or} \ |C\overset{\cdot}{\pm}N|) \quad \rightarrow \quad |N\equiv C-C\equiv N|$$

there is an increase of at least one in the total number of bonds, and therefore cyanogen is much more stable than the two CN radicals. As for ClO_2, the most favorable form of this molecule according to Green and Linnett [24] should be described as $|\overline{O} = \underline{C}l \overset{\cdot}{-} \underline{O}|$ and $|\overline{O} \overset{\cdot}{-} \underline{C}l = \overline{O}|$ for in this structure the bond order is higher than in other structures. Hence, one of the ClO_2 bonds is a three-electron bond, where one electron appears to be associated primarily with the chlorine atom, one primarily with the oxygen, and one in the region between them, i.e., in the bond. According to this structure, the ClO_2 molecule has 3.5 bonds, and two molecules have 7 bonds. If a dimer formed, it would also have 7 bonds, and therefore dimerization in this case is not favorable. Apparently, for the same reason, chlorine dioxide does not tend to react with organic molecules in the characteristic free radical pathways. Instead, it tends to react with both inorganic and organic compounds as a one electron accepter, being thereby reduced to chlorite.

Although ClO_2 does not break s bonds to produce new ones, it can easily react with organic molecules containing lone pair electrons or π electrons which are susceptible to oxidation. The organic molecule which loses an electron by such oxidation is thus converted into a free radical which, in turn, can either combine with another molecule of ClO_2 to produce a chlorite ester, or can be stabilized by some intramolecular rearrangements.

A more detailed description of these reactions is given elsewhere in this chapter, with regard to the organic reactions of chlorine dioxide.

3
Inorganic Reactions of Chlorine Dioxide

3.1
Redox Reactions

The major redox reactions of chlorine dioxide are those in which it acts as a one electron acceptor. However, in the presence of stronger oxidizers, it may itself be oxidized to chlorate. One can easily deduce thermodynamically possible oxidations involving chlorine dioxide from the list of Redox potentials shown in Table 1.

Apart from standard potentials, redox potentials at pH 7.0 have also been calculated and introduced in Table 1 in order to give a better reflection of the situation under actual water works conditions.

As one can see from Table 1, chlorine dioxide is a weaker oxidant than ozone, hypochlorous acid or chlorine, but stronger than hypochlorite, especialy at pH 7.0, and substantially stronger than iodine. Its oxidation potential is little lower than that of the $Br_2/2\,Br^-$ couple and therefore at the concentrations prevailing in actual water plants chlorine dioxide does not oxidize bromide to bromine. This has an important significance for the disinfection of drinking water rich in bromide. When such water is disinfected with either ozone or chlorine, the bromide is oxidized into bromine and hypobromous acid, which in turn reacts with naturally occurring aquatic humic substances to produce undesirable brominated compounds including THM (mainly bromoform) [27, 69]. This does not occur with chlorine dioxide disinfection.

In solutions of neutral pH, iodide is oxidized by chlorine dioxide to iodine, while ClO_2 is reduced to chlorite, according to Eq. (1):

$$2\,ClO_2 + 2\,I^- \quad \rightarrow \quad 2\,ClO_2^- + I_2 \qquad (1)$$

Table 1. Redox Potentials at 25 °C [22, 25, 26]

Reaction	$E°$ (volts)	$E^{pH=7}$ (volts)
$O_3\,(g) + 2H^+ + 2e^- \rightleftharpoons O_2\,(g) + H_2O$	2.0	1.59
$HClO + H^+ + 2e^- \rightleftharpoons Cl^- + H_2O$	1.49	1.28
$Cl^2\,(aq) + 2e^- \rightleftharpoons 2Cl^-$	1.40	1.40
$2ClO_3^- + 4H^+ + 2e^- \rightleftharpoons 2ClO_2 + 2H_2O$	1.15	0.32
$Br_2 + 2e^- \rightleftharpoons 2Br^-$	1.09	1.09
$ClO_2 + e^- \rightleftharpoons ClO_2^-$	0.95	0.95
$ClO^- + H_2O + 2e^- \rightleftharpoons Cl^- + 2OH^-$	0.90	0.49
$NO_3^- + 2H+ + 2e^- \rightleftharpoons NO_2^- + H_2$	0.84	0.43
$ClO_2^- + 4H^+ + 4e^- \rightleftharpoons Cl^- + 2H_2O$	0.78	0.37
$O_2 + 2H^+ + 2e^- \rightleftharpoons H_2O_2$	0.68	0.27
$I_2 + 2e^- \rightleftharpoons 2I^-$	0.62	0.62
$SO_4^- + 9H^+ + 8e^- \rightleftharpoons HS^- + 4H_2O$	0.25	−0.21
$CNO^- + H_2O + 2e^- \rightleftharpoons CN^- + 2OH^-$	−0.97	−1.38

This reaction proceeds completely to the right and is the basis for the standard iodometric method for the quantitative determination of chlorine dioxide. In acidic media below pH 4 (but not at pH 7.0) chlorous acid may further react with iodide, being itself converted into chloride according to Eq. (2):

$$2HClO_2 + 8I^- + 6H^+ \rightleftharpoons 4I_2 + 2Cl^- + 4H_2O \qquad (2)$$

Therefore one may often find in the literature the overall Eq. (3) describing the oxidation of iodide by ClO_2 in acidic media:

$$2ClO_2 + 10\,I^- + 6H^+ \rightleftharpoons 5I_2 + 2Cl^- + 4H_2O \qquad (3)$$

However, it should be noted that this is merely a combination of Eqs. (1) and (2), while the actual mechanism is more complicated and involves several steps.

Chlorine, having a higher oxidation potential than chlorine dioxide, may oxidize ClO_2 according to Eq. (4):

$$2ClO_2 + Cl_2 + 2H_2O \rightleftharpoons 2ClO_3^- + 2Cl^- + 4H^+ \qquad (4)$$

Interestingly, there is considerable data in the literature on the reverse reaction, in which ClO_2 is formed [29–32], while the information on the forward reaction is very limited. Using radio-labeled $_{17}Cl^{38}$, Taube and Dodgen [13] found that it is the ClO_2 which is oxidized to chlorate, while the Cl_2 is reduced to Cl^-. In spite of the substantial difference in oxidation potentials between Cl_2 and ClO_2, which indicates an equilibrium in favor of the forward reaction (to the right), its reaction rate is very slow. At 18 °C, in a solution containing 0.1 mol/l ClO_2, 0.1 mol/l Cl_2, $5 \cdot 10^{-3}$ mol/l acid and $2 \cdot 10^{-3}$ mol/l chloride, only a small percent of ClO_2 had reacted in 18 days [15].

However, in neutral solutions, where Cl_2 is largely disproportioned into chloride and hypochlorous acid, the oxidation of chlorine dioxide by hypochlorous acid proceeds much faster, according to the reaction described in Eq. (5):

$$2ClO_2 + HClO + H_2O \rightleftharpoons 2ClO_3^- + Cl- + 3H^+ \qquad (5)$$

This reaction is accompanied by a disproportionation of ClO_2, and therefore the overall rate law proposed for its kinetics was [33]:

$$-\frac{d[HClO]}{dt} = K_1[ClO_2][HClO] + K_2[ClO_2] \tag{6}$$

where at 25 °C: $K_1 = 1.28$ mol^{-1} l s^{-1} ; $K_2 = 0.022$ min^{-1}.

This reaction may explain the presence of chlorate, which is often found in drinking water disinfected with combinations of chlorine dioxide and chlorine, when chlorine is introduced some time after the chlorine dioxide. (Such combinations are often used in order to reduce the formation of both THM produced by chlorine, and chlorite produced by chlorine dioxide.)

According to Table 1 the oxidation potential of the couple NO_3^-/NO_2^- at pH 7.0 is substantially lower than of ClO_2, and therefore under neutral conditions nitrite can be easily oxidized by chlorine dioxide according to Eq. (7):

$$NO_2^- + 2\,ClO_2 + H_2O \rightleftharpoons NO_3^- + 2\,ClO_2^- + 2\,H^+ \tag{7}$$

The reaction rate was found to be of the first order with respect to each reactant and second order overall, with the second order rate constant being 113 (± 3) mol^{-1} l s^{-1} [58]. This has an important significance for water treatment, because nitrite is often found in water as a metabolite of the denitrification process. The concentrations of nitrite in such waters can sometimes deplete under natural conditions most of the chlorine dioxide used.

Other important species oxidized by ClO_2 are cyanide, sulfide and sulfite, which are oxidized according to the following overall (not mechanistic) reactions:

$$2\,ClO_2 + CN^- + 2\,OH^- \longrightarrow CNO^- + 2\,ClO_2^- + H_2O \tag{8}$$

$$2\,ClO_2 + 2\,Na_2S \longrightarrow 2\,NaCl + Na_2SO_4 + S \tag{9}$$

$$2\,ClO_2 + 5\,H_2SO_3 + H_2O \longrightarrow 5\,H_2SO_4 + 2\,HCl \tag{10}$$

Cyanide in complexes with Zn or Cd is also oxidized by ClO_2 to cyanate, especially at pH > 10, with the stoichiometric molar ratio $ClO_2/CN^- = 5.2$. The cyanide complex of Ni is also oxidized, but with the above ratio being 1.0 [34].

Most metals like Cd, Zn, Mg, and Al in the powdered state are oxidized by chlorine dioxide to the corresponding chlorites, by reactions similar to that described for Zn in Eq. (11):

$$2\,ClO_2 + Zn\ dust\ (in\ H_2O) \longrightarrow Zn(ClO_2)_2 \tag{11}$$

Some chlorites are not stable and are hydrolyzed to the respective hydroxides.

3.2
Disproportionations

Both chlorous acid and chlorine dioxide may disproportionate in acidic solutions, mainly below pH$_2$, according to Eq. (12):

$$4\,HClO_2 \longrightarrow Cl^- + 2\,ClO_2 + ClO_3^- + 2\,H^+ + H_2O \tag{12}$$

The rate of this reaction increases as the pH is decreased and the temperature increases. However, at very high temperatures, near the boiling point of water, another disproportionation may also take place according to Eq. (13) [35]:

$$5\,HClO_2 \longrightarrow 4\,ClO_2 + HCl + 2\,H_2O \tag{13}$$

The reaction in Eq. (13) may be significant even at ordinary temperatures (at low pH) when the chloride concentration is high, because under such circumstances the rate of the competitive reaction described in Eq. (12) decreases due to the common ion effect of the chloride.

Chlorine dioxide is stable at ordinary temperature under neutral conditions and disproportionates only when the pH is increased above 11. In strongly basic solutions ClO_2 disproportionates according to Eq. (14):

$$2\,ClO_2 + 2\,OH^- \longrightarrow ClO_2^- + ClO_3^- + H_2O \tag{14}$$

Although this reaction appears simple, its mechanism is actually very complicated and includes bimolecular and monomolecular steps, for which the following kinetic equation has been proposed [22, 36, 37]:

$$-\frac{d\,[ClO_2]}{dt} = K_1[ClO_2] + K_2[ClO_2]^2 \tag{15}$$

At pH 12, K_1 was found to be $0.3 \cdot 10^{-2}\,min^{-1}$, and K_2, $7.1\,mol^{-1}\,min^{-1}$.

The bimolecular step predominates at the beginning of the reaction, when the ClO_2 concentration is high, while the monomolecular step predominates toward the end, when the ClO_2 concentration decreases. For the low concentrations actually used in water works, $t_{1/2}$ as determined from K_1 is 3.8 h, and this is indeed the $t_{1/2}$ measured at pH 12 in concentrations of $3-10\,mg\,l^{-1}\,ClO_2$.

3.3
Health Effects

An overview of the inorganic reactions of chlorine dioxide reveals that the main inorganic product of ClO_2 is chlorite and, to a lesser extent, chlorate.

In actual water works disinfected by chlorine dioxide (either drinking water or wastewater effluents), it was found that 40–50% of the chlorine dioxide consumed was converted into chlorite [38]. This result can be rationalized based on the nature of chlorine dioxide organic reactions, which are described later in this chapter. Data regarding chlorate production in actual water plants is incomplete, but it is estimated that 10% of the chlorine dioxide consumed is converted into chlorate.

As early as 1937, Richardson showed that both chlorite and chlorate may cause methemoglobinemia [39]. Therefore, when chlorine dioxide was first introduced in the late 1970s as an alternative disinfectant for drinking water, there was much concern about the health risks associated with the production chlorite and chlorate, as the main inorganic by-products of chlorine dioxide usage. This concern stimulated intensive biological investigations which clarified that it is the depletion of glutathione (GSH) which caused hemolytic anemia rather than the direct oxidation of hemoglobin. It seems that chlorite catalyzes the formation of hydrogen peroxide within the cell, which is responsible for the decrease in glutathione concentrations, and that interference with glutathione peroxidase may mediate the critical effect.

Heffernan and coworkers in 1979 showed a dose-related decrease in erythrocyte glutathione in cats that were administered with 50 mg/l of chlorite in drinking water for 30 days [8]. A similar effect was observed by Couri and co-workers [9, 40] in rats and chicken that were exposed to concentrations as low as 10 mg/l chlorite in drinking water for a period of 12 months. Apart from that, by using laboratory animals, it has been established that 10 mg/l of chlorite in drinking water may inhibit ^{3}H-thymidine incorporation into nuclei of liver cells [40]. Red cell glutathiodione depression was also observed in animals exposed to chlorine dioxide, but in that case the glutathione depression disappeared with continued exposure (after 4 months). This kind of adaptivity was not found for chlorite [40].

Therefore, the main concern was of neonates and people sensitive to oxidative damage to hemoglobin, like those suffering from glucose-6-phosphate dehydrogenase (G6PD) deficiency.

A controlled clinical evaluation of chlorine dioxide, chlorite and chlorate that was carried out for 3 months on a group of more than a hundred volunteers, including a few G6PD deficient people, failed to demonstrate any undesirable clinical signals in any of the participating subjects, who received 500 ml water with 5 mg/l oxidants daily [41].

In 1981 an epidemiological study regarding the health effects of ClO_2-treated drinking water was summarized by Michael et al. [42]. That epidemiological study, which included clinical pathology tests, was conducted on the 198 inhabitants of a small Appalachian village in the USA, which was supplied with chlorinated water during the winter months, and chlorine dioxide-treated water during the summer. Statistical analysis of the data failed to identify any significant exposure related effects.

A clinical study of the blood chemistry parameters of 20 renal dialysis patients that had been drinking ClO_2 (1 mg/l)-treated drinking water for 12 months also did not show evidence of ClO_2 induced anemia, or of any other adverse health effects [43].

Many other clinical evaluations with volunteers and epidemiologic studies that have been carried out in recent years have also failed to demonstrate any adverse health risks associated with the consumption of drinking water treated with ClO_2 in concentrations which are ordinarily applied [41, 44, 45].

Although none of the above-mentioned toxicological data have revealed any adverse health effects to people consuming drinking water treated with ClO_2 at concentrations of 1 mg/l or less, Bull and Kopfler [82] suggested that the MCL (Maximum Contamination Level) for ClO_2 in regulatory actions should not exceed 0.1 – 0.2 mg/l. They based their suggestion on the assumption that the most critical effect of ClO_2 may be a delayed brain development observed in two strains of rats, that were exposed to a dose of 14 mg/kg/day of ClO_2 in drinking water [83, 84].

Assuming that a young child of 10 kg consumes 1 l/day of drinking water, and the EPA ordinarily applies under such circumstances a safety factor of 1000, the MCL was calculated as

$$\frac{14 \text{ mg/kg/day} \cdot 10 \text{ kg}}{1 \text{ l/day} \cdot 1000} = 0.14 \text{ mg/l}$$

As for chlorite and chlorate, Bull and Kopfler [82] suggested MCL of 0.034 mg/l and 0.0034 mg/l respectively, based on the hematological effects discussed previously.

Practically speaking, in light of the fact that the residual concentration of ClO_2 is usually substantially lower than the initial concentration (as a result of water demand) the meaning of an MCL value of 0.2 mg/l ClO_2 for ordinary water consuming 0.3–0.5 mg/l ClO_2 is that an initial (treatment) concentration of 0.5–0.7 mg/l may be appropriate. However, it will be much harder to meet the low standard recommended by Bull and Kopfler for chlorite. Therefore the WHO guidline, according to which the standard for chlorite should be around 0.2 mg/l, seems much more realistic.

4
Organic Reactions of Chlorine Dioxide

4.1
A General Introduction to the Organic Reactions

Until approximately 15 years ago, when chlorine dioxide was introduced as an alternative disinfectant for drinking water, most of the information concerning organic reactions of chlorine dioxide came from the pulp and paper bleaching industries, and from other studies that were performed under conditions that were vastly different from those encountered in water treatment plants. Many side-reactions were possible in those systems, leading to products that would not be produced under normal conditions for drinking water treatment. The discussion in this chapter is limited to those reactions, mechanisms and products that may occur at neutral or nearly neutral pH conditions and ambient temperatures. Other reactions are discussed in previous reviews available in this field [15, 22, 46–48].

The fact that the ClO_2 molecule has an odd number of electrons led early investigators to propose free radical pathways for its reactions with many organic compounds. The most common proposed mechanism was hydrogen abstraction. However, most currently available data supports an electron transfer (E.T.) mechanism, according to which ClO_2 acts as a one-electron acceptor in most reactions. The E.T. process can be well rationalized in light of the molecular structure of ClO_2, as described previously in this chapter. This in no way excludes hydrogen abstraction, which still may occur either concurrently with E.T. or even exclusively in certain cases, especially when a very energetic activated hydrogen is available. Nevertheless, if a definite mechanism is not known for a reaction, E.T. is assumed to be the most likely mechanism since it can explain many known reactions of ClO_2 and predict most of their products.

4.2
Chlorine Dioxide Reactions with Amines and Amino Acids

Amines were the first organic group for which the reactions with chlorine dioxide were investigated intensively, including kinetics, mechanisms and products, in a series of ten papers published by Rosenblatt and co-workers (49a–i).

The reactivity of amines, as reported in those studies, was in the order tertiary > secondary > primary.

Ammonia does not react with chlorine dioxide at all, but triethylamine reacts with ClO_2 ten times as fast as phenol, and nearly 10^4 times as fast as the amine reacts with permanganate.

The kinetic equation of the reaction of ClO_2 with most aliphatic secondary and tertiary amines is as shown in Eq. (16):

$$-\frac{d[ClO_2]}{dt} = K_2[ClO_2][amine] \tag{16}$$

The reaction is thus first order with respect to each reactant and second order overall. The concentration of amine in Eq. (16) is the free base concentration. For triethylamine at 25 °C, $K_2 = 2.15 \cdot 10^5$ mol^{-1} l s^{-1}.

However, when the stoichiometry and material balance of the reaction were investigated, it was found that two moles of ClO_2 are required for each mole of amine, as shown for example in Eq. (17) for triethylamine:

$$Et_3N + 2\,ClO_2 + H_2O \;\rightarrow\; Et_2NH + CH_3CHO + 2\,ClO_2^- + 2\,H^+ \tag{17}$$

It was also found that the reaction was retarded by the addition of chlorite ion.

Based on the above observations, Hull et al. [49c] proposed the following mechanism (Eq. 18):

$$
\begin{aligned}
&ClO_2 + RCH_2N{\overset{R'}{\underset{R'}{\diagup\kern-0.4em\diagdown}}} \;\overset{5\,LOW}{\rightleftharpoons}\; ClO_2^- + RCH_2\overset{+}{N}{\cdot}{\overset{R'}{\underset{R'}{\diagup\kern-0.4em\diagdown}}} \quad (I)\\[1em]
&I \;\rightarrow\; H^+ + RCH{\cdots}N{\overset{R'}{\underset{R'}{\diagup\kern-0.4em\diagdown}}} \quad (II)\\[1em]
&II + ClO_2 \;\overset{fast}{\rightarrow}\; RCH{=}\overset{+}{N}{\overset{R'}{\underset{R'}{\diagup\kern-0.4em\diagdown}}} + ClO_2^- \quad (III)\\[1em]
&III \;\xrightarrow[-H^+]{H_2O}\; RCHO + HN{\overset{R'}{\underset{R'}{\diagup\kern-0.4em\diagdown}}}
\end{aligned}
\tag{18}
$$

According to this mechanism the rate-determining step is a reversible one-electron oxidation of the amine by ClO_2. An aminium cation-radical (I) is thus produced, stabilized by the repulsion of an α-proton. This mechanism finally leads to a carbon-nitrogen cleavage with the formation of an aldehyde and a lower amine (a tertiary amine is converted into a secondary amine).

The alternative hydrogen abstraction mechanism (shown in Eq. 19) was ruled out by the following observations:

$$ClO_2 + RCH_2N\underset{R'}{\overset{R'}{<}} \xrightarrow{slow} RCH \cdots N\underset{R'}{\overset{R'}{<}} + HClO_2 \qquad (19)$$
$$\Updownarrow$$
$$H^+ + ClO_2^-$$

a) The reaction was retarded by chlorite even at a pH much above the pK_a of chlorous acid, when the acid was completely ionized, while according to Eq. (19) only the undissociated chlorous acid could retard the reaction.

b) In the case of benzyldimethylamine and its substituted derivatives, the product distributions showed an equal probability for either a benzyl or a methyl cleavage, while according to the hydrogen abstraction mechanism the phenyl activation of α-carbon would lead to a preponderant attack at the position a to the phenyl, which would result in preferential benzyl cleavage.

c) When a deuterated trimethylamine-d_9 was reacted with ClO_2 the measured isotope effect was 1.3. An α-hydrogen abstraction as the rate determining step according to Eq. (19) would give rise to a primary isotope effect with $K_H/K_D > 4$.

In accordance with the E.T. mechanism, the reaction rates, on a logarithmic scale for a series of p- and m-substituted benzyldimethylamines were directly proportional to the pK_a values of the amines, giving rise to a linear Bronsted plot with a slope of 0.81 and a correlation coefficient of 0.99. The higher the basicity of the amine and the electron density on its nitrogen, the faster its reaction with chlorine dioxide. A few example reaction rates are shown in Table 2.

While the E.T. mechanism is general for most aliphatic tertiary and secondary amines, the case may be different for benzylamine, which is a primary amine possessing an activated α-hydrogen adjacent to the phenyl group. In this

Table 2. Reaction rates for some p- and m-substituted benzyldimethylamines with ClO_2 at 26 °C [49b]

Substituent	pK_a	K_2 ($mol\,l^{-1}\,sec^{-1}$)
p-methoxy	9.32	$4.94 \pm 0.78 \cdot 10^4$
p-methyl	9.22	$3.52 \pm 0.27 \cdot 10^4$
unsubstituted	9.03	$2.74 \pm 0.24 \cdot 10^4$
p-chloro	8.82	$1.99 \pm 0.21 \cdot 10^4$
m-chloro	8.67	$1.57 \pm 0.14 \cdot 10^4$
p-nitro	8.14	$4.48 \pm 0.17 \cdot 10^3$

case, hydrogen abstraction may be facilitated, and indeed the reaction of benzylamine with ClO_2 is not retarded by chlorite. Therefore it was concluded that in this case the hydrogen abstraction process predominates.

One should also bear in mind that these reactions are very sensitive to tereochemical and electronic effects within the molecule, a fact that may sometimes dictate modified or alternative mechanistic pathways leading to different products. Thus, the reaction rate of triethylendiamine, which is actually a cyclic β-amino acid, with chlorine dioxide is second order with respect to ClO_2, with a proposed mechanism as shown in Eq. (20) [49h]:

$$:N \diagdown\diagup N: + ClO_2 \rightleftharpoons :N \diagdown\diagup \overset{+}{N} + ClO_2^-$$

$$I \qquad\qquad\qquad II$$

$$II + ClO_2 \xrightarrow[\text{step}]{\text{rate determining}} \overset{+}{N} \diagdown\diagup \overset{+}{N} + ClO_2^- \qquad (20)$$

$$III$$

$$III \xrightarrow{\text{fragmentation}} CH_2{=}^+N \qquad N^+{=}CH_2 \xrightarrow{H_2O} HN \qquad NH + 2CH_2O + 2H^+$$

$$IV$$
$$\text{piperazine}$$

This reaction results in a β-carbon-carbon cleavage, rather than the usual carbon-nitrogen cleavage shown in Eq. (18).

A similar cyclic amine which does not have a second amino group in the β position is quinuclidine (V). For quinuclidine, the β-cleavage shown above is not possible, and the regular mechanism for aliphatic amines shown in Eq. (18) is not possible either, because structure II in Eq. (18) implies a planar configuration which is stereochemically impossible in this case [49h]. Therefore this molecule reacts (very slowly) by a different mechanism which leads to the formation of the N-oxide [50]:

$$\underset{V}{\diagup\!\!\!\diagdown N} \xrightarrow{ClO_2} N \text{ oxide} \qquad (21)$$

The formation of carbinol amines, rather than carbon-nitrogen cleavage, often occurs in the oxidation of tertiary amines contained in five- or six-membered rings. An example is the oxidation of N-butylisoindoline (VI) by ClO_2 to produce N-butyl-3-hydroxyphthalimidine (VII) shown in Eq. (22) [46]:

$$\underset{VI}{\overset{CH_2}{\underset{CH_2}{\diagup\!\!\!\diagdown}}N} \xrightarrow{ClO_2} \underset{VII}{\overset{\overset{O}{\|}}{\underset{CH_2}{\diagup C \diagdown NH}}} \qquad (22)$$

Although the detailed mechanistic pathways for all these reactions are beyond the scope of this chapter, it should be noted that most of them proceed via the electron transfer process in the first step. The aminium cation radical then reacts further in a different way. Subseqent e.s.r. studies indicated that most aromatic amines also react with ClO_2 via the same initial step [53, 54].

In amino acids, the amino group is not basic, and therefore it does not react with ClO_2. Noss and co-workers [51] applied chlorine dioxide to 19 amino acids, of which only 6 reacted. Those were proline, histidine, cystein, tyrosine, tryptophan and methionine. All these amino acids possess reactive groups (other than NH_2) which would react even if they were not incorporated within an amino acid. An example shown in Eq. (23) is the oxidation of the phenolic group in tyrosine to the quinonin group in dopaquinone:

$$\underset{\text{tyrosine}}{\underset{\overset{|}{OH}}{\overset{\overset{\displaystyle CH_2-CH-COOH}{|}}{\bigcirc} \; NH_2}} \quad \xrightarrow{ClO_2} \quad \underset{\text{dopaquinone}}{O=\bigcirc=CH-CH-\underset{\overset{|}{NH_2}}{COOH}} \tag{23}$$

Dopaquinone is the colored substance produced in the colorimetric determination of chlorine dioxide with tyrosine [55].

4.3
Chlorine Dioxide Reactions with Phenols

Like the reactions of chlorine dioxide with amines, its reactions with phenols are also second order, first order with respect to both ClO_2 and phenol, according to Eq. (24) [56]:

$$-\frac{dt[ClO_2]}{dt} = 2\,K_2[Ph][ClO_2] \tag{24}$$

The factor 2 in Eq. (24) derives from the reaction stoichiometry, according to which ClO_2 disappears twice as fast as the phenol (see Eq. 29 below).

However, since the protonated and unprotonated forms of phenol have different reaction rates, and at neutral pH both forms are present, Eq. (24) should be split in order to reflect the reaction rates of both species. If $[Ph]_T$ is the total concentration of phenol (including both forms), the concentrations of the protonated and unprotonated forms can be expressed by Eqs. (25) and (26) respectively, where K_a is the dissociation constant of phenol

$$[PhOH] = \frac{[Ph]_T \cdot H^+}{K_a + H^+} \tag{25}$$

$$[PhO^-] = \frac{[Ph]_T \cdot K_a}{K_a + H^+} \tag{26}$$

Substituting Eqs. (25) and (26) into the overall kinetic at Eq. (27) gives rise to Eq. (28), where K_{PhOH} and K_{KPhO-} are the second order rate constants of the protonated and unprotonated phenol species, respectively:

$$-2\frac{d[ClO_2]}{dt} = \{K_{PhOH}[PhOH] + K_{PhO-}[PhO^-]\} \times [ClO_2] \tag{27}$$

$$-\frac{d[ClO_2]}{dt} = \frac{2(K_{PhOH}H^+ + K_{PhO-} - K_a)}{K_a + H^+}[Ph]_T[ClO_2] \tag{28}$$

The values for K_{PhOH} and K_{PhO-} as determined by Wajon et al. [56], are $0.24(\pm0.01)$ mol^{-1} l s^{-1} and $2.4(\pm0.2) \cdot 10^7$ mol^{-1} l s^{-1}, respectively. These values differ from the original values obtained by Grimley and Gordon [57] only by the stoichiometric factor 2, which in the original paper was included in the second order rate constant. Very similar results to those obtained by Grimley and Gordon have recently been reported by Hoignè and Bader [58] who obtained the values $0.4(\pm0.1)$ mol^{-1}l s^{-1} and $4.9(\pm0.5) \cdot 10^7$ mol^{-1}l s^{-1} for K_{PhOH} and K_{PhO-}, respectively.

The stoichiometry and material balance for the reaction of phenol with ClO_2 (when the latter was initially present as a three-fold excess) as determined by Wajon et al. [56] are shown in Eq. (29):

$$\text{phenol} + 2\,ClO_2 \rightarrow \text{benzoquinone} + ClO_2^- + ClO^- + 2\,H^+ \tag{29}$$

Based on the kinetics and stoichiometry of Eq. (29), Wajon and co-workers [56] proposed the mechanism shown in Eq. (30) for the reaction of phenol with ClO_2:

According to this mechanism, the rate-determining step is the electron transfer from the phenoxide anion to ClO_2 (very similar to the rate determining step in the reaction between ClO_2 and amine). The phenoxide radical (I_a) is stabilized by combination with a second ClO_2 molecule. This is followed by the release of hypochlorous acid, giving rise to benzoquinone as the final product. A very similar mechanism was proposed for the reaction of ClO_2 with hydroquinone (for which the kinetics are the same), except for the fact that the first, slow, oxidation process is followed by a second fast oxidation step of the phenoxide radical to a biradical which is transformed to benzoquinone [56].

Consistent with the mechanism described above, a good correlation was found between the oxidation rates of substituted phenoxide anions and their respective σ^- Hammett constants according to Eq. (31), giving rise to a ϱ value of $-3.2(\pm 0.4)$ [59, 60]. The negative ϱ value is expected for an electron transfer mechanism, which is favored by increased electron density at the reaction center [61]:

$$\log K = \log K_0 + \varrho\sigma^- \tag{31}$$

K_0 is the rate constant of the unsubstituted phenoxide anion.

Rate constants for oxidation of some substituted phenols and phenoxide anions by ClO_2 are shown in Table 3.

The observation that the ClO_2 oxidation rate constants for phenoxide anions are about 10^6 times those of the respective protonated phenols can be attributed to two factors:

1) the redox potentials of phenols are considerably higher than those of phenoxide anions (the latter can therefore be oxidized easier);
2) the proton dissociation, which is a necessary step in phenol oxidation (see Eq. 30), slows down the reaction.

Table 3. Second order rate constants for the reactions of ClO_2 with some substituted phenols and phenoxide anions

Substituent	$2 k_{PhOH}$ (mol^{-1} l s^{-1})	$2 k_{PhO^-}$ (mol^{-1} l s^{-1})	Ref.
4-hydroxy (hydroquinone)	$3.9 \cdot 10^4$	$6.5 \cdot 10^9$	[56]
4-methoxy	$2.5(\pm 0.2) \cdot 10^4$	$1.7(\pm 0.6) \cdot 10^9$	[58]
		$1.48 \cdot 10^9$	
3-methoxy		$4.9 \cdot 10^7$	[60]
4-methoxy	$5.0(\pm 0.1) \cdot 10^1$	$5.2(\pm 0.3) \cdot 10^8$	[58]
		$5.2 \cdot 10^8$	[60]
unsubstituted phenol	$0.4(\pm 0.1)$	$4.9(\pm 0.5) \cdot 10^7$	[58]
	$0.48(\pm 0.02)$	$4.8(\pm 0.4) \cdot 10^7$	[56]
3-methyl		$1.0(\pm 0.1) \cdot 10^8$	[58]
		$0.94 \cdot 10^8$	[60]
4-bromo		$5.4 \cdot 10^7$	[60]
4-nitro	$1.4(\pm 0.4) \cdot 10^1$	$4.0(\pm 0.1) \cdot 10^3$	[58]

Oxidation rates of other substituted phenoxides, for which experimental data is not available yet, can be predicted by the quantitative structure activity relationship (QSAR) which was recently established by Tratnyek and Hoign [59] The QSAR is based on the good correlation found between the oxidation rate constants (on a logarithmic basis) and several molecular free energy indicators, as follows:

a) the Hammett σ-constants;
b) the polarographic half-wave oxidation potentials ($E_{1/2}$);
c) the Marcus equation for outer-sphere electron transfer reactions [62].

With regard to products, it has already been noted that at a molar ratio ClO_2/phenol = 3.5 the main, and perhaps only, product of phenol oxidation is 1,4-benzoquinone. Additional products may appear either when phenol is in excess, or when the chlorine dioxide concentration is larger than 3.5 times the concentration of phenol. For instance, when phenol was in excess, Spanggord detected several chlorinated phenols including 2-chlorophenol,4-chlorophenol, 2,6-dichlorophenol, 2,4-dichlorophenol and 2,4,6-trichlorophenol [63]. The chlorophenol production can be attributed to the hypochlorous acid released upon the reaction of ClO_2 with phenol (Eq. 30d). This hypothesis is supported by the observation that the rate of production of chlorophenols by ClO_2 is very similar to their production by hypochlorous acid. In addition, the distribution of chlorophenols produced by ClO_2 is similar to that found when phenol and HClO are reacted under similar conditions. On the other hand, when ClO_2 is in a large excess, further oxidation may lead to the formation of maleic and oxalic acids. A general scheme describing all possible products between ClO_2 and phenol, as proposed by Masschelein [22], is shown in Fig. 1. Glabisz [64] suggests that the dichlorobenzoquinones produced arise by way of the corre-

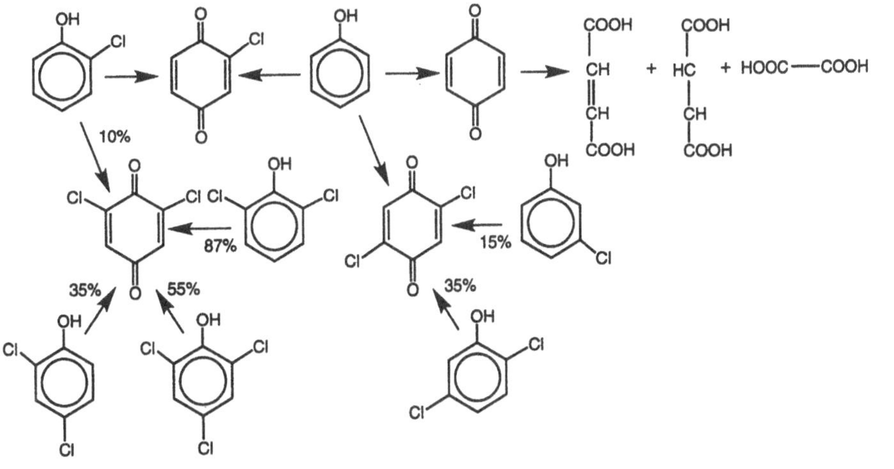

Fig. 1. Reaction scheme for chlorine dioxide with phenol and chlorophenols [22]

sponding chlorophenols, and this conclusion is apparently justified, since a) 2,4,6-trichlorophenol is oxidized by ClO_2 to 2,6-dichloro-1,4-benzoquinone with a 55% yield [64], and b) 1,4-benzoquinone does not react with ClO_2 [65].

Some of the more detailed investigations concerning the reaction products of ClO_2 with phenols have been carried out by Glabisz, who has made some generalizations on the reaction types [64]. According to Glabisz, the phenols may be divided into two groups. The first group includes monohydric phenols that are not para-substituted, and hydroquinone. They retain their ring structure upon reaction with ClO_2 and are oxidized mainly to quinones and chloro-quinones. The second group consists of monohydric phenols that are para-sub-stituted, such as p-cresol, 3,4-xylenol, and di- or tri-hydric phenols with two hydroxyl groups in the ortho or meta- positions, such as resorcinol, pyrocatechol or pyrogallol. These undergo ring cleavage upon reaction with ClO_2 and are oxi-dized to organic acids (oxalic, maleic and fumaric), and to carbon dioxide.

However, this generalization does not always appear to be correct, and many exceptions can be found. Thus, 2,4-dichlorophenol is easily oxidized by ClO_2 to give 2,6-dichloro-1,4-benzoquinone.

4-Hydroxybenzaldehyde is oxidized rapidly to 1,4-benzoquinone by chlorine dioxide at pH 6 and below, but the yield is only 21–27% [15]. By contrast, vanillin undergoes ring scission between the hydroxy and methoxy groups to give β-formylmuconic acid monomethyl ester [66]:

$$(32)$$

Vanillin alcohol is oxidized at low pH to give two products, one that results from the previously described displacement, and the other characteristic of ring scission [15]:

$$(33)$$

Substituted hydroquinones are easily oxidized as expected, to the corresponding quinones. Some examples are given below:

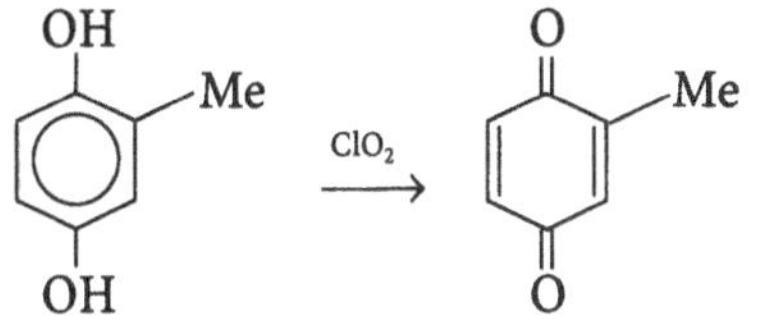

(34)

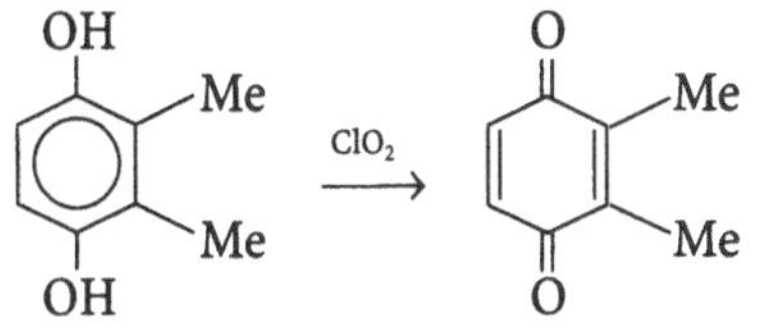

(35)

From the toxicological point of view, the most significant of all the above products are the benzoquinones found in all investigations of the reactions of chlorine dioxide with phenolic derivatives. Very little is known about the toxicity of these quinones. Recently short-term bacterial bioassays using *pseudomonal fluorescence* have been conducted on benzoquinone and hydroquinone [67]. The growth of the bacteria was strongly inhibited by benzoquinone at concentrations ranging from 10–200 ppm. Hydroquinone was less inhibitory to the bacterial growth. However, the oxygen uptake of the bacteria was strongly diminished by as little as 10 ppm of either chemical.

4.4
Reactions with Alkenes

Unlike the reactions of ClO_2 with amines and phenols which are very rapid, with second order rate constants of $4.3 \cdot 10^3 - 2 \cdot 10^5 \text{ mol}^{-1} \text{l}^{-1} \text{s}^{-1}$ for amines, and of $0.5 - 4.0 \cdot 10^4 \text{ mol}^{-1} \text{l}^{-1} \text{s}^{-1}$ for undissociated phenols, the reaction rates with alkenes are relatively slow with $k_2 = 3 \cdot 10^{-3} - 1.5 \text{ mol}^{-1} \text{l}^{-1} \text{s}^{-1}$. Thus, the long time required for such reactions can be relevant only for water works with long distribution lines and many extensions, where water may be retained for hours.

The reactions of chlorine dioxide with olefins have been studied by many investigators, and nearly as many mechanisms have been suggested for these reactions. However, due to the fact that every investigator used different model compounds for his studies it is hard to draw critical conclusions with regard to the right mechanistic pathways. It is also possible that several pathways may proceed either concurrently or even simultaneously, depending on the molecular structure of the reactant, its chemical affinities, and the reaction conditions.

Lindgren et al. [70] proposed an hydrogen abstraction mechanism for the aqueous reactions of ClO_2 with cyclohexene and methyl oleate (Eq. 36), based on the reaction products which for cyclohexene (in a seven-fold excess) were mainly cyclohex-1-ene-3-one (I, 13%), 3-chlorocyclohex-1-ene (II, 14%) and *trans*-2-chlorocycloxan-1-ol (15%). Other products that were recovered in much smaller yields were 2-chlorocyclohexan-1-one (7%), *trans*-1,2-dichlorocyclohexane (7%) and adipic acid (4%).

(36)

A similar mechanism was attributed by Lindgren and Svahn [71] to the reaction of ClO_2 with methyl oleate, the products of which were of the same type. However, Rav-Acha and coworkers, who studied the reactions of ClO_2 with allylbenzene [72], indene [73] and styrene [74, 75], indicated that there are many olefins for which in the first step the hydrogen abstraction process is very unlikely. Instead, they proposed an electron transfer (E.T.) mechanism, according to which the alkene is oxidized to a cation-radical, and ClO_2 is reduced to chlorite according to Eq. (37):

(37)

This mechanism is very similar to that assigned previously for the reactions of ClO_2 with amines and phenols. Rav-Acha and coworkers based their arguments on the following observations:

1) the isotope effect (k_H/k_D) found for the reaction of ClO_2 with 1,1,3-trideuterated indene (in both H_2O and D_2O) was 1.0, thus refuting the hydrogen abstraction process for which the expected isotope effect would be 5.8–6.9 [76, 77];
2) a good correlation was found between the reactivity of olefins towards ClO_2 and their ionization potentials, as shown in Table 4; it was also shown that the order of reactivities shown in Table 4 is in accord with the stabilities of the cation-radicals formed at the transition states of those reactions [73];
3) a strong solvent effect was observed when ClO_2-olefin reactions were carried out in water-isopropanol mixtures, with a decreasing reaction rate for in-

Table 4. Pseudo-first order rate constants for various olefins ($2 \cdot 10^{-5}$ mol l^{-1}) with excess ClO_2 ($9 \cdot 10^{-4}$ mol l^{-1}) at pH 7.0, and their ionization potentials (I.P.)

Alkene	k_{abs} (h^{-1})	I.P. (e.v.)
indene	4.5	8.2
α-methylstyrene	1.5	8.5
styrene	0.2	8.7
cyclohexene	0.07	9.2
allylbenzene	0.008	–
cinnamic acid	unreactive	–

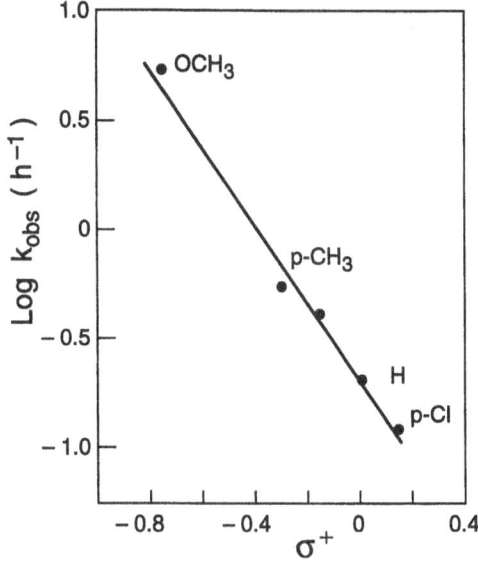

Fig. 2. Hammett plot for the reaction of ClO_2 with p-substituted styrens

creasing isopropanol fraction [73], thus indicating transition states which are more polar than the starting materials;

4) as one can observe from Fig. 2, a linear Hammett plot has been obtained between the rate constants for the reactions of ClO_2 ($9 \cdot 10^{-4}$ mol l^{-1}) with p-substituted styrenes ($5 \cdot 10^{-5}$ mol l^{-1}) and their σ^+ values, giving rise to a negative ϱ value of -1.45, which indicates that a positive charge is generated on the carbon atom at the reaction center.

(The fact that the ϱ value was still lower than with reactions where classical carbonium ions are formed, was attributed to the relatively low activation energy required for the formation of a cation-radical (15 Kcal) as compared to 23 Kcal required for the formation of classical carbonium ions [78]).

In order to encompass the data presented above, the reactions of ClO_2 with indene and allylbenzene, which are among the most and least reactive olefins, respectively, were intensively investigated including kinetics, stoichiometry and reaction products. Thus, the kinetics for reaction of ClO_2 with indene at 25 °C were found to be of the first order with respect to each reactant, and second order overall, with k_2 of 131 mol^{-1} l min^{-1} at pH 7.0, 155 mol^{-1} l min^{-1} at pH 5.0, and 170 mol^{-1} l min^{-1} at pH 4.0. The stoichiometry has also altered with pH. At pH 7.0 the ClO_2: indene ratio was 2:1 and the chlorite produced accounted for 25% of the ClO_2 consumed. At pH 4.0 the ClO_2:indene ratio was 1:1 and no chlorite was detected. The change in stoichiometry was attributed to disproportionation of chlorite at pH 4.0 according to Eq. (38):

$$2\,HClO_2 \rightarrow HClO + H^+ + ClO_3^- \tag{38}$$

The hypochlorous acid produced reacts further with a second molecule of indene, thus giving rise to a ClO_2:indene ratio of 1:1 instead of 2:1 at pH 7.0. The products recovered for the ClO_2:indene reaction at pH 4.0 were *trans*-1,2-indanediol (I, 14%), *cis*-1,2-indanediol (II, 23%), 1-hydroxyindan-2-one (III, 14%), *cis*-2-chloroindan-1-ol (IV, 23%) and *trans*-2-chloroindan-1-ol (V, 26%). Based on the data presented above the proposed mechanism for the reaction is shown in Fig. 3.

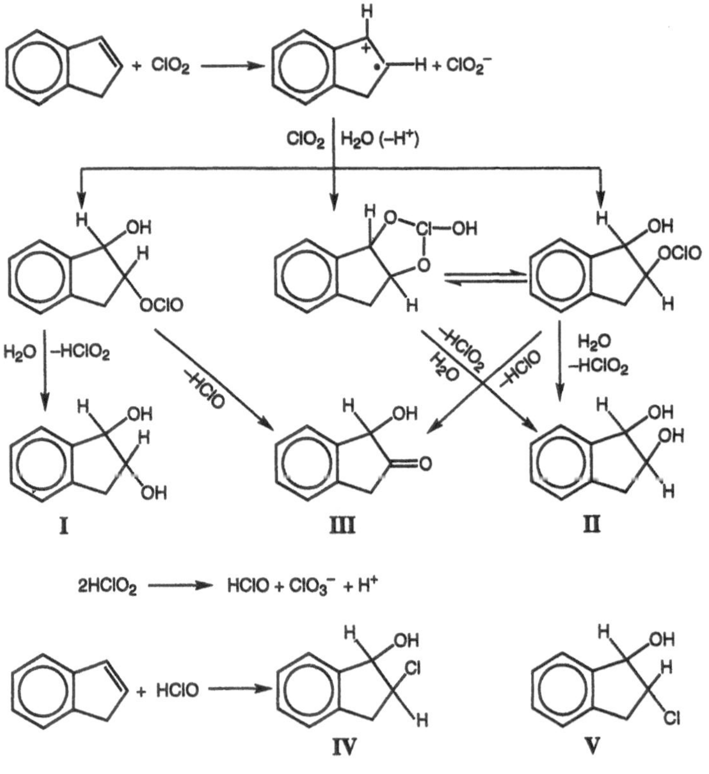

Fig. 3. Proposed mechanism for the reaction of ClO_2 with indene at pH 4.0

The ClO_2-indene reaction is thus regio-specific leading to the preponderance of the *cis*-diol (II) over the *trans*-diol (I), and to the formation of 1-hydroxy-2-indanone (III) rather than the more common 2-hydroxy-1-indanone. All of those observations can be rationalized in terms of the mechanism presented above.

The overall stoichiometry and material balance for the reaction of indene with ClO_2 at pH 4.0 is shown in Eq. (39):

$$2 \text{ indene} + 2 \text{ } ClO_2 + 2 \text{ } H_2O \rightarrow \text{indandiol} + \text{chlorohydrin}$$
$$+ H^+ + ClO_3^- \tag{39}$$

A similar pathway was proposed by Rav-Acha et al. [72] for the reaction of ClO_2 with allylbenzene at pH 4.0, which led to the formation of 3-phenyl-1-chloropropan-1-ol (7%), 3-phenyl-1-chloropropan-2-ol (20%), cinnamaldehyde (8%), cinnamic acid (8%) and benzoic acid (5%) (Fig. 4). 52% of the initial material remained unchanged after 48 h.

In a similar manner Rav-Acha and coworkers [79] explained the products recovered by Kolar and Lindgren [80] for the reaction of ClO_2 with styrene. However, Kolar and Lindgren proposed a free radical type mechanism for this

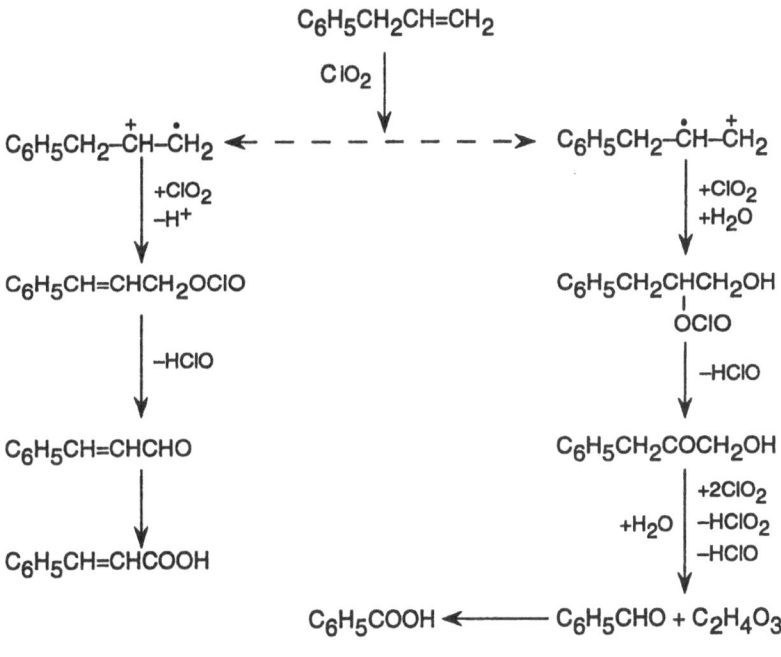

Fig. 4. Proposed pathways for the reaction of ClO_2 with allylbenzene at pH 4.0

reaction which would involve ClO$^•$ radicals, according to Eq. (40). As shown in Eq. (40) the main products of the reaction at pH 4.0 were 2-chloro-1-phenyl-1-ethanol (III, 36%) and 1-phenyl-1,2-ethanediol (II, 27%). At pH 6.0 the main difference was that styrene oxide (phenyl-oxirane, I) was recovered in 26% yield, while the yield of diol (II) was reduced to 6%:

$$C_6H_5-CH=CH_2 + ClO_2 \quad \rightarrow \quad \overset{O}{\overset{/\backslash}{C_6H_5-CH-CH_2}} + ClO$$
$$\text{I}$$

$$\text{I} + H_2O \quad \rightarrow \quad \underset{\underset{\text{II}}{\overset{|}{OH}\ \overset{|}{OH}}}{C_6H_5-CH-CH_2} \tag{40}$$

$$ClO + ClO_2 + H_2O \quad \rightarrow \quad HClO + ClO_3^- + H^+$$
$$C_6H_5-CH=CH_2 + HClO \quad \rightarrow \quad \underset{\overset{|}{OH}\ \overset{|}{Cl}}{C_6H_5-CH-CH_2}$$
$$\text{III}$$

Likewise, Lindgren and Nilsson [81] explained the products for the reaction of ClO$_2$ with stilbene and its derivatives in carbon tetrachloride solutions, that for *trans*-stilbene were α-chlorobenzyl phenyl ketone (43%), *trans-$\alpha\alpha'$*-epoxy-bibenzyl (36%) and $\alpha\alpha'$-dichlorobibenzyl (7%), accompanied by a small amount of benzyl. Of course, the same products could also have been explained by the E.T. process described previously.

4.5
Reactions with Polycyclic Aromatic Hydrocarbons (PAH)

Although detailed mechanistic studies regarding the reactions of chlorine dioxide with PAH have not yet been carried out, there are several indications that these reactions proceed via a one-electron oxidation process at least at their rate-determining step, similar to the reactions of ClO$_2$ with phenols, amines and some olefins that were described previously. Thus, it was found that the sequence of ClO$_2$ reaction with PAH is in accord and with the reactivities of the regions which are the most sensitive towards oxidation [85]. These regions were defined by Pullman and Pullman [86] as the L-regions similar to the 9,10 position of anthracene. This is in contrast to the sequence of chlorine reactivities that reflects better the reactivities of the K-regions which are the most reactive double-bonds toward electrophilic attack. In addition, Table 5 shows that the reactivities of PAH toward ClO$_2$ follow both, the polarographic half-wave oxidation potentials of these PAH, and their ionization potentials as calculated by Dewar and Lepley [87] based on Huckel's method [85]. Actually it is the χ index as calculated by Dewer and Lepley, which is proportional to the ionization potential (I.P.) that is presented in Table 5, rather than the ionization potential itself.

Unlike the reactions of PAH with chlorine, the ClO$_2$ reactions are not affected by pH changes, or the presence of light, but are affected by the presence

Table 5. The sequence of ClO_2 reactions with PAH [85]

PAH	$t_{1/2}$ (min)	$\chi^{(87)}$	$E_{1/2}$ (V vs SCE)
Benzo(a)pyrene	~0.1	0.371	0.94
Anthracene	0.15	0.414	1.09
Benzo(a)anthracene	1	0.452	1.18
Benzo(e)pyrene	200	0.497	1.25
1,2,7,8-Dibenzanthracene	600	0.503	1.26
Chrysene	2100	0.520	1.35
Fluoranthene	–	0.618	1.45
Naphthalene	–	0.618	1.45

of air (headspace) or oxygen. Many of the ClO_2 reactions were found to be of the second order, first with respect to ClO_2 and first with respect to PAH [85].

It seems that, in contrast to chlorine which may react with PAH in a variety of mechanisms including addition, substitution or oxidation processes, chlorine dioxide reacts mainly as an electron acceptor. Therefore its reactions are much more specific and selective. Many PAH that react quite easily with Cl_2 do not react, or react very slowly, with ClO_2 (e.g., naphthalene or fluoranthene), while other PAH, such as benzo(a)pyrene (B(a)P), anthracene and benzo(a)-anthracene (B(a)A), react with ClO_2 a great deal more rapidly than with Cl_2. Interestingly, the most carcinogenic PAH members B(a)P and B(a)A react with ClO_2 much faster than with Cl_2 to produce less toxic compounds [88].

Pyrene can be used as an appropriate model in order to demonstrate the difference between the reactions of Cl_2 and ClO_2, as Cl_2 reacts mainly at the 4,5 or 9,10 positions of pyrene, while ClO_2 reacts exclusively at its 1,6 or 1,8 positions which are the more sensitive to oxidation. A possible pathway for the reaction of ClO_2 with pyrene, as postulated by Likkonen et al. [89] is shown in Fig. 5.

As shown in Fig. 5, the 1,6- and 1,8-pyrenediols were the major products of the reaction of ClO_2 ($3-4 \cdot 10^{-3}$ mol l^{-1}) with pyrene at either pH 3 or pH 8 [89]. These major products were accompanied by minor quantities of mono- and di-chloro-pyrenes. However, for some PAH such as anthracene, fluorene and benzo(a)pyrene the quinonoid compounds are the predominating or exclusive products. Thus, fluorene and anthracene are converted by ClO_2 exclusively to 9-fluorenone and 9,10-anthraquinone respectively [89]. For benzo(a)pyrene a few quinonoid-type products accounted together for more than 90% of the yield. The three quinonoid-type products of benzo(a)pyrene are 3,4-benzopyrene-1,5-quinone, 3,4-benzopyrene-5,8-quinone, and 3,4-benzopyrene-5,10-quinone [90]. The remaining 10% of the reaction yield contained small amounts of chlorinated derivatives. 1-Methylphenanthrene with ClO_2 produces, at pH 8.0, 1-methyl-phenanthrene-9,10-quinone together with mono- and di-chlorinated derivatives [89]. The chlorinated products may be attributed either to impurities of chlorine, or to chlorine which may be produced as a by-product during the organic reactions of ClO_2 (as was explained previously

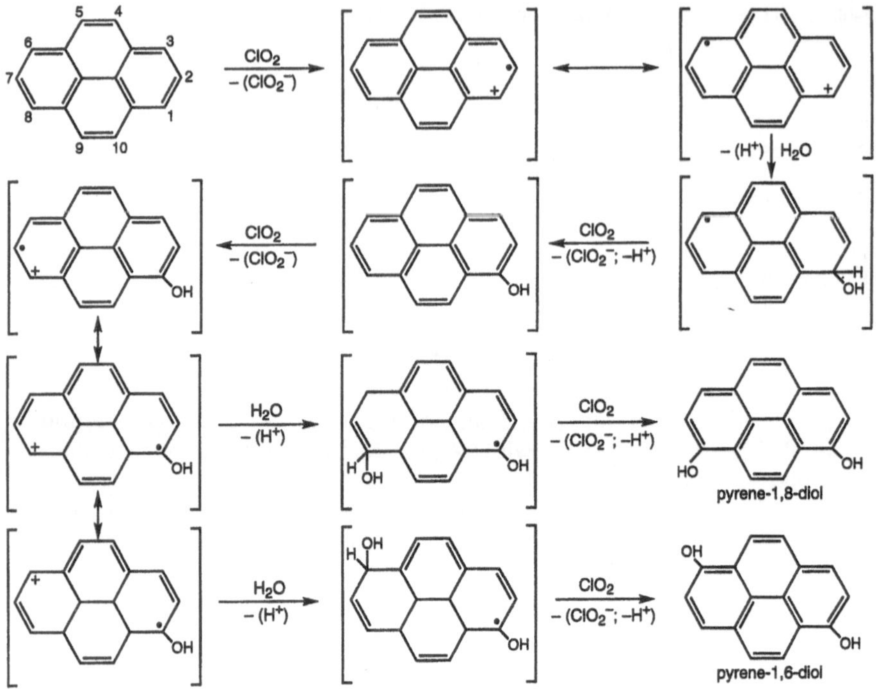

Fig. 5. A possible mechanistic pathway for the reaction of ClO$_2$ with pyrene [89]

with respect to the ClO$_2$ reactions of olefins). Chlorine may also be produced in the same way by the disproportionation of chlorite in presence of organic compounds.

4.6
Reactions of Chlorine Dioxide with Hydrocarbons, Alcohols, Carbonyl Compounds and Carboxylic Acids

Most aliphatic and aromatic hydrocarbons not containing specific reactive functional groups, do not react with chlorine dioxide under regular water treatment conditions [93]. However, some polycyclic aromatic hydrocarbons, which are easily oxidized, do react with ClO$_2$ as discussed previously in this chapter.

Carboxylic acids are likewise unreactive with respect to chlorine dioxide unless they contain specific reactive groups. Paluch et al. [94], however, reported that benzylic acid may produce small amounts of benzoic acid upon its reaction with aquatic ClO$_2$ under neutral pH conditions.

Alcohols are also resistant to ClO$_2$ at neutral pH, even at elevated temperatures. Only under extreme conditions which include acidic media, high temperature and excess ClO$_2$, do alcohols react to produce aldehydes and carboxylic acids. Thus, at pH 1 and 70–80 °C, ethanol and 2,3-butandiol are converted into acetic acid, most probably via the diacetyl intermediate [95]. Glucose at pH

2.0 is oxidized at the -CH$_2$OH groups to produce carbonyl and carboxylic groups, apparently without a ring rupture [96].

Under similar conditions (pH 3.0, T = 50 °C, excess ClO$_2$), cellulose produces the following acids: gluconic, arabonic, erythronic, glyoxalic and oxalic [97]. Carbonyl compounds react quite readily with ClO$_2$ even under moderate conditions to produce carboxylic acids.

Otto and Paluch (98) reported that benzaldehyde reacted violently with ClO$_2$. Other aldehydes and ketones, such as acetaldehyde, n-butyraldehyde and diacetyl, have also been found to react with ClO$_2$ at neutral pH and moderate temperatures [95]. Still it should be asked, however, to what extent these reactions actually occur in drinking water plants. Stevens [10] was able to identify many aldehydes in Ohio River water treated with chlorine dioxide. This suggests that they were not further oxidized spontaneously to their respective carboxylic acids. Hoigné and Bader [99] have also reported that benzaldehyde was unreactive towards 0.1 ppm of ClO$_2$. However, it seems that at least some aldehydes can be oxidized to carboxylic acids under water treatment conditions, because carboxylic acids have been recently identified among the characteristic by-products of chlorine dioxide disinfection [29].

4.7
Inactivation of Pesticides by Chlorine Dioxide

Pesticides are often found in water plants as a consequence of run-off from treated agricultural areas. Chemical disinfectants sometime oxidize pesticides present in water, thereby purifying the water. However, some pesticides may be converted into even more toxic compounds through partial oxidation.

Organo-chlorine pesticides such as lindane, DDT and toxaphene are unaffected by either Cl$_2$ or ClO$_2$ in concentrations used during water treatment [100, 101]. Most carbamate insecticides, like aldicarb, methomyl, carbaryl and propoxur are also unreactive towards ClO$_2$ at neutral pH in the concentration range of 0-10 ppm [102]. These include insecticides (like aldicarb and methomyl) that contain C-S bonds (which are very reactive towards ClO$_2$ in the case of amino acids), although these insecticides react easily with chlorine or ozone [102]. The widely used herbicides paraquat and diquat are also unreactive towards ClO$_2$.

However, there are pesticides that react with ClO$_2$ even faster than with chlorine. Thus, 100 ppb of rotenone can be removed from water within 15 min by 0.8 ppm ClO$_2$, while 45 ppm of Cl$_2$ are required to achieve the same result [103]. Parathion at neutral pH is also oxidized by ClO$_2$ much faster than by chlorine, but in the latter case the oxidation is not necessarily an advantage because the oxidation product, paraoxon, is more toxic than the parent compound parathione. This problem does not exist at elevated pH levels, because above pH 9 paraoxon is spontaneously hydrolysed into p-nitrophenol [104]:

Parathion → (ClO₂) → Paraoxon → O_2N—⟨⟩—OH → Degredation products

5
Mutagenic Assessment

In this chapter it is shown that chorine dioxide may produce a variety of by-products by reacting with aquatic humic substances or with other organic compounds which may be present in water. The toxicological features of these products are virtually unknown. Therefore, general toxicological assays are used to obtain some evaluation of the possible long-term adverse health effects associated with the disinfection of water by ClO_2. One of the most popular general assessments is the Ames *Salmonella*/microsome assay for mutagenicity [105, 106]. This assay is based on a mutant of *Salmonella typhimurium* which is unable to synthesize histidine. When this bacteria is applied to a petri-dish containing a nutrient agar devoid of histidine, colonies will not develop unless the bacteria is exposed to a back mutation-causing mutagen. Several strains of such bacteria have been developed. These strains are either sensitive to different groups of mutagens, or indicate different DNA changes, such as base substitution or frame shift [105–107]. Kool et al. [108] reviewed the mutagenic properties of disinfected water according to the Ames assay, and concluded that 5–15 mg l^{-1} of ClO_2 may increase the direct mutagenicity of water. The increased mutagenicity is most probably caused by the formation of mutagens which are different from those produced by chlorine, and are more pronounced without S 9 metabolic activation. However, generally the increase in mutagenic activity by ClO_2 is significantly lower than that obtained by chlorine. Following the assumption that the mutagenicity is related to the organic content of water, most investigators used methods for separation and concentration (like extraction, resin adsorption, or lyophilization) [108–110], in which only the organic compounds could be collected and assessed for mutagenicity, disregarding the disinfectant itself and its inorganic by-products (chlorite and chlorate).

As for chlorine dioxide and its inorganic by-products, there are controversial opinions in the literature regarding their mutagenicity. The problems connected with testing these chemicals for mutagenicity in in vitro systems arise from the fact that these are biocides especially designed to kill bacteria and cells by damaging the cell membrane. Those chemicals may thus directly injure or kill the test cells, rendering the results invalid. It should be noted, however, that by the time the water is consumed the concentrations of these chemicals are relatively low.

Following the assumption that the main precursors for mutagens are the naturally occurring aquatic humic substances, several investigators examined aqueous mixtures composed of humic substances and disinfectants for mutagenicity [111, 112]. Both commercial humic acid and natural water-isolated humic substances were used for those studies. The commercial acid was not mutagenic, but a dose-related increase in mutagenicity was observed upon treatment with either chlorine or chlorine dioxide. However the humic solutions with ClO_2 were generally less mutagenic than with chlorine. The mutagenicity of water-isolated humic substances varied according to the source of water. In some cases the humic substances themselves were slightly mutagenic, and chlorination did not alter the mutagenicity significantly. Mutagenicity was slightly reduced when treated with ClO_2, althoug some cytotoxic effects were also observed.

6
Summary

The redeeming feature of chlorine dioxide is the selectiveness of its reactions, as opposed to those of chlorine. As a consequence it is not as much consumed by water and does not produce as many by-products. This in itself can be regarded as an advantage, because by lessening the number of by-products produced, the possibility of formation of toxic products lessens. However, ClO_2 is also not an ideal oxidant (that would theoretically produce only CO_2) and as such it produces a variety of by-products, some of which have been described throughout this chapter. Apart from the concern for the undesirable inorganic by-products chlorite and chlorate that was previously discussed, the main concern is about the formation of aldehydes that were already identified in actual water plants disinfected by chlorine dioxide, and quinones or benzoquinones that may be produced upon the reaction of ClO_2 with phenols which are very frequently found in surface water.

As a result of the health concern about disinfection by-products it has now become widely accepted that the best way to minimize the formation of possible toxic by-products is to eliminate their precursors by using advanced water treatment procedures (which would generally include coagulation-flocculation, filtration, GAC adsorption, and disinfection). However, when and where such advanced treatment is not feasible, various combinations of disinfectants are sometimes used in order to reduce the concentration of one undesirable by-product or another. The combinations most frequently used are those of chlorine and chlorine dioxide, chlorine dioxide and monochloramine (NH_2Cl), or chlorine dioxide and ozone. A combination of ClO_2 and Cl_2 may reduce the concentrations of both THM produced by chlorine, and chlorite produced by ClO_2 [113, 114]. When chlorine dioxide is applied some time before the chlorine, it may react with the precursors for THM (probably humic materials) to reduce the formation of THM upon chlorination. The chlorine, in turn, oxidizes the chlorite (produced by ClO_2) back to chlorine dioxide or to chlorate (although chlorate is also undesirable). When a combination of chlorine dioxide and monochloramine is used, the ClO_2 is applied first in order to achieve a satisfying disinfection, while monochloramine is applied subsequently to maintain a residual disinfection over a prolonged period.

From the point of view of research there is an urgent need for more research regarding ClO_2 by-products which are produced in actual drinking water plants in order to evaluate better the long-term health effects associated with ClO_2 disinfection. This is because most of our present knowledge is obtained from laboratory experiments with model compounds, which do not necessarily reflect the situation in actual drinking water.

Acknowledgment. The author wishes to thank Professor Mark M. Benjamin from University of Washington, Seattle WA for revising the manuscript.

References

1. Rook JJ (1974) Water Treat Exam 23:234
2. Bull RJ (1980) J Am Water Works Assoc 72:299
3. Cantor KP, Hoover R, Mason TJ, McCabe LJ (1978) J Natn Cancer Inst 61:979
4. Augenstein HW (1971) J Am Water Works Assoc 66:716
5. Longley KW, Moore BE, Sorber CW (1980) J Water Pollut Control Fed 52:2098
6. Cronir S, Scarpiro PV, Zink ML (1980) In: Jolley RL, Gorchev H, Hamilton DH (eds) Water chlorination: environmental impact and health effects. Ann Arbor Science, Ann Arbor MI., vol 2, pp 651–659
7. Sussman QS, Ward U (1981) Engineering aspects of chlorine dioxide. Proceedings of Seminar on Control of Organic Chemical Contaminants in Drinking Water. USEPA, Washington DC
8. Heffernan WP, Guion C, Bull RJ (1979) J Environ Pathol Toxicol 2:1487
9. Couri D, Abdel-Rahman MD (1980) J Environ Pathol Toxicol 3:451
10. Stevens AA (1982) Environ Health Presp 46:101
11. Rosenblatt DH (1978) In: Rice RG, Cotruvo JA (eds) Ozone/chlorine dioxide oxidation products of organic materials. Ozone Press International, Cleveland Ohio, pp 332–343
12. Eméleus HJ, Sharpe AG (eds) (1963) Advances in inorganic and radiochemistry. Academic Press, NY, vol 5, p 52
13. Taube H, Dodgen J (1949) J Amer Chem Soc 71:3330
14. White JF (1949) In: Kirk PE, Othmer DF (eds) Encyclopedia of chemical technology, vol 3. Interscience Encyclopedia, NY, pp 696–707
15. Gordon G, Kieffer RG, Rosenblatt DH (1972) Prog Inorg Chem 15:201
16. Brett RW, Ridgeway JW (1981) J Inst Water Eng Sci 35:135
17. Bowen EJ, Cheung WM (1932) J Chem Soc. 1200
18. Bray WC (1906) Z Physik Chem 54:569
19. Williamson HV, Hampel CA (1954) US Patent No 2 683 651
20. Clark AH, Bengley B (1970) J Chem Soc 46(A):460
21. Clark AH (1971) J Mol Struct 7:485
22. Masschelein WJ (1979) Chlorine dioxide: chemistry and environmental impact of oxychlorine compounds. Ann Arbor Science, Ann Arbor, MI
23. Nielsen AH, Woltz PJH (1952) J Chem Phys 20:1878
24. Green M, Linnett JW (1960) J Chem Soc: 4959
25. Latimer W (1952) Oxidation potentials, 2nd edn. Prentice-Hall, Englewood Cliffs, NJ
26. Stumm W, Morgan JJ (1970) Aquatic chemistry. Wiley Interscience, NY
27. Rav-Acha Ch, Choshen E, Serri A, Limoni B (1983) Environ Sci Health 18:651
28. Rav-Acha Ch, Choshen E, Serri A, Limoni B (1985) Environ Poll 10:47
29. Beuermann L (1965) Gas-Wasserfach 106:783
30. Keating E (1952) Papier 9:155
31. Luther K, MacDougall F (1906) Z Physik Chem 55:477

32. Luther K, MacDougall F (1908) Z Physik Chem 62:1908
33. Flis IE, Mishchenko KP, Salvis KYu (1959) Zh Prikl Khim 32:284
34. Belevtzev AN, Maksimenko YuL (1975) In: USA/USSR Symp Phys Chem, USEPA, Washington DC, pp 105–110
35. Barnett B (1935) The decomposition of chlorous acid. Ph.D. Dissertation. University of California, LA
36. Goodeve CF, Richardson FV (1937) J Amer Chem Soc 59:294
37. Gordon G, Feldman F (1964) Inorg Chem 3:1728
38. Limoni B, Chosen E, Rav-Acha Ch (1984) Environ Sci Health 19:943
39. Richardson AP (1937) J. Pharmacol Exper Therap 59:101
40. Abdel-Rahman MS, Couri D, Bull RJ (1981) J Environ Pathol Toxicol 5:867
41. Lubbers JR, Chauhan S, Bianchine JR (1982) Environ Health Presp 46:57
42. Michael GE, Miday RK, Berez JP, Miller RJ, Kraemer OF, Lucas JB (1981) Arch Environ Health 36:20
43. Ames RG, Stratton JW (1987) Arch Environ Health 42:280
44. Lubbers JR, Chauhan S, Miller JK, Bianchine JR (1984) J Environ Pathol Toxicol Oncol 5:229
45. Lubbers JR, Bianchine JR (1984) J Environ Pathol Toxicol Oncol 5:215
46. Miller GW, Rice RG, Robson CM, Kühn W, Wolf H (1978) USEPA Report 600|012-78-147, Cincinati, OH
47. Stevens AA, Seeger DR, Slocam CJ (1978) In: Rice RG, Cotruvo JA (eds) Ozone/ClO₂ oxidation products of organic materials. Intl Ozone Inst, Clevelend, OH, pp 332–343
48. Rav-Acha Ch (1984) Water Res 18:1329
49. (a) Rozenblatt DH, Hayes AJ, Harrison BL, Streaty RA, Moore KA (1963) J Org Chem 28:2790; (b) Rozenblatt DH, Hull LA, De Luca DC, Davis GT, Weglein RC, Williams KR (1967) J Amer Chem Soc 89:1158; (c) Hull LA, Davis GT, Rozenblatt DH, Williams KR, Weglein RC (1967) J Amer Chem Soc 89:1163; (d) Dennis WH, Hull LA, Rozenblatt DH (1967) J Org Chem 32:3783; (e) Rozenblatt DH, Davis GT, Hull LA, Forberg GD (1968) J Org Chem 33:1649; (f) Davis GT, Rozenblatt DH (1968) Tetrahedron Lett 4085; (g) Hull LA, Davis GT, Rozenblatt DH (1969) J Amer Chem Soc 91:6247; (h) Hull LA, Davis GT, Rozenblatt DH, Mann CK (1969) J Phys Chem 73:2142; (i) Hull LA, Giordano WP, Rozenblatt DH, Davis GT, Mann CK, Milliken SB (1969) J Phys Chem 73:2147
50. Burrows EP, Rozenblatt DH, (1982) J Org Chem 47:892
51. Noss CI, Hauchman FS, Olivieri VP (1985) Wat Res 20:351
52. Hodgen HW, Ingols RS (1954) Anal Chem 26:1224
53. Ozawa T, Kwan T (1983) Chem Pharm Bull 31:2864
54. Ozawa T, Kwan T (1984) Chem Pharm Bull 32:1589
55. Neta P, Huie RE, Ross AB (1988) J Phys Chem Ref Data 17:1213, 1229
56. Wajon JE, Rosenblatt DH, Burrows EP (1982) Environ Sci Technol 16:396
57. Grimley E, Gordon G (1973) J Inorg Nucl Chem 35:2388
58. Hoigné, Bader H (1994) Wat Res 28(1):45
59. Tratnyek PG, Hoigné J (1994) Wat Res 28:57
60. Alfassi ZB, Huie RE, Neta P (1986) J Phys Chem 90:4156
61. Exner O (1988) Correlation analysis of chemical data. Plenum Press, NY
62. Eberson L (1987) Electron transfer reactions in organic chemistry. Springer, Berlin Heidelberg New York
63. Spanggord RJ (1978) Private communication. Stanford Research Inst, Stanford, CA
64. Glabisz U (1968) The reactions of chlorine dioxide with components of phenolic wastewater – Summary. Monograph 44, Polytechnic University, Szcecin, Poland
65. Dodgen H, Taube H (1949) J Amer Chem Soc 71:2501
66. Sarkanen KV, Kakehi K, Murphy RA, White H (1962) Tappi 45:24
67. Trevors JT, Bursuraba J (1980) Bull Environ Contam Toxicol 25:672
68. Rook JJ (1978) J Environ Sci Health A13:91
69. von Gunten U, Hoigné J (1992) J Water SRT – Aqua 41:299

70. Lindgren BO, Svahn CM, Widmark G (1965) Acta Chem Scand 19:7
71. Lindgren BO, Svahn CM (1966) Acta Chem Scand 20:211
72. Rav-Acha Ch, Choshen E, Sarel S (1986) Helv Chim Acta 69:1728
73. Rav-Acha Ch, Choshen E (1987) Environ Sci Technol 21:1069
74. Choshen E, Blitz R, Rav-Acha Ch (1986) Tetrahed Lett 49:5989
75. Rav-Acha Ch, Choshen E, Blitz R, Grafstein O (1989) In: Jolley RL, Bull RJ, Davis WP, Katz S, Roberts MH, Jacobs VA (eds) Water chlorination: chemistry, environmental impact and health effects, vol 6. Lewis, Chelsea, MI, pp 849–858
76. Bergson G (1964) Acta Chem Scand 18:2003
77. Hine J (1962) Physical organic chemistry, 2nd edn. McGraw-Hill, Tokyo, p 72
78. Noyce DS, King PA, Lane CA, Reed WL (1962) J Amer Chem Soc 84:1678
79. McManus SP (1973) Organic reactive intermediates. Academic Press, NY
80. Kolar JJ, Lindgren BO (1982) Acta Chem Scand 36:599
81. Lindgren BO, Nilsson T (1974) Acta Chem Scand 28:847
82. Bull RJ, Kopfler FC (1991) Health effects of disinfectants and disinfection by-products. Monograph, American Water Works Association Research Foundation, Denver, CO
83. Taylor DH, Pfohl RJ (1985) In: Jolley RL, Bull RJ, Davis WP, Katz S, Roberts MH, Jacobs VA (eds) Water chlorination: chemistry, environmental impact and health effects, vol 5. Lewis, Chelsea MI, pp 355–364
84. Toth GP, Long RE, Smith MK (1990) J Toxicol Environ Health 31:29
85. Rav-Acha Ch, Blitz R (1984) Wat Res 19:1273
86. Pullman A, Pullman B (1955) Adv Cancer Res 3:117
87. Dewar MJS, Lepley AR (1961) J Am Chem Soc 83:4560
88. Richert JK, Kuute H, Engelhardt K, Bornett J (1971) Arch Hyg Bakt 155:18
89. Likkonen RJ, Lin S, Oyler AR, Lukasewyes MT, Cox DA, Yu ZJ, Carlson RM (1983) In: Jolley RL, Brungs WA, Cumming RB (eds) Water chlorination: chemistry, environmental impact and health effects, vol 4. Ann Arbor Science, Ann Arbor MI, pp 151–165
90. Richert JK (1968) Arch Hyg Bakt 152:265
91. de Greef E, Morris JC, van Kreijl CF, Morra CFH (1980) In: Jolley RL, Brungs WA, Cumming RB (eds) Water chlorination: chemistry, environmental impact and health effects, vol 3. Ann Arbor Science, Ann Arbor MI, pp 913–924
92. Richardson SD, Thruston AD Jr, Collette TW, Patterson KS, Lykins Jr BW, Majetich G, Zhang Y (1994) Environ Sci Technol 28:592
93. Stevens AA, Seeger RD, Slocum CJ (1978) In: Rice RG, Cotruvo JA (eds) Ozone/chlorine dioxide oxidation products of organic materials. Ozone Press International, Cleveland OH, pp 383–399
94. Paluch K, Otto J, Kozlowski K (1965) Rocznicki Chem 39:1603
95. Somsen RA (1960) Tappi 43:154
96. Flis IE (1955) Tr Leninger Tech Int 16:62
97. Becker ES, Hamilton JK, Lucke WE (1965) Tappi 48:60
98. Otto J, Paluch K (1965) Rocznicki Chem 39:171
99. Hoigné J, Bader H (1982) Vom Wasser 59:253
100. Buescher CA, Douyherty JH, Skrinde RT (1964) J Water Pollut Control Fed 36.1005
101. Robeck GG, Dostal KA, Cohen JM, Kreiss l JF (1965) J Am Water Works Assoc 57:181
102. Mason (Zelicovitz) Y, Choshen E, Rav-Acha Ch (1990) Wat Res 24:11
103. Cohen JM, Kamphake LJ, Lemake PE, Henerson C, Woodward RL (1960) J Am Water Works Assoc 52:1551
104. Gomaa HM, Faust SD (1972) In: Gould RF (ed) Advances in chemistry series, vol. 111. American Chem Soc, Washington DC
105. Ames BN, Yanofsky C (1971) In: Hollaender A (ed) Chemical mutagens: principles and methods for their detection, vol 1. Plenum Press, NY, pp 267–282
106. Ames BN, McCann J, Yamasaki E (1975) Mutat Res 31:347
107. Levin DE, Hollstein M, Christman MF, Schwiers EA, Ames BN (1982) Proc Natl Acad Sci USA 79:7445

108. Kool HJ, van Kreijl CF, Hrubec J (1985) In: Jolley RL, Bull RJ, Davis WP, Katz S, Roberts MH, Jacobs VA (eds) Water chlorination: chemistry, environmental impact and health effects, vol 5. Lewis, Chelsea MI, pp 187–205
109. Kool HJ, van Kreijl CF, van Kranen HJ, de Greef E (1981) Chemosphere 10:85
110. Zoeteman BCJ, Hrubec J, de Greef E, Kool HJ (1982) Environ Health Presp 46: 197
111. Guttman-Bass N, Bairey-Albuquerque M, Ulitzur S, Chartrand A, Rav-Acha Ch (1987) Environ Sci Technol 21:252
112. Meier JR, Lingg RD, Bull RJ (1983) Mutat Res 118:25
113. Rav-Acha Ch, Serri A, Choshen E, Limoni B (1984) Wat Sci Tech 17:611
114. Noach MG, Doerr RL (1978) In Jolley RL, Gorchev H, Hamilton DH Jr (eds) Water chlorination: chemistry, environmental impact and health effects, vol 2. Ann Arbor Science, Ann Arbor MI, pp 49–58

Subject Index

Environmental Chemistry

Volume 5 / B

J. Hrubec (Ed.)

Quality and Treatment of Drinking Water

1995. IX, 166 pp. 57 figs., 17 tabs.
Hardcover DM 218,-
ISBN 3-540-58178-2

The concern over the entry of agrochemicals and other xenobiotics into drinking water resources and over the general quality of drinking water is increasing. The topic of water quality and water supply will continue to be of great interest during the next two decades in developed as well as in developing countries. The new volume discusses in an authoritative way the key issues of drinking water and its often necessary treatment.

Please order from
Springer-Verlag Berlin
Fax: + 49 / 30 / 8 27 87- 301
e-mail: orders@springer.de
or through your bookseller

Price subject to change without notice.
In EU countries the local VAT is effective.

Springer

The Handbook of

Environmental Chemistry

Volume 3 / I

A.H. Neilson (Ed.)

PAHs and Related Compounds

Chemistry

1997. XXIV, 422 pp.
168 figs., 23 tabs.
Hardcover DM 298,–
ISBN 3-540-62394-9

Volume 3 / J

A.H. Neilson (Ed.)

PAHs and Related Compounds

Biology

1997. XXIV, 386 pp.
96 figs., 38 tabs.
Hardcover DM 298,-
ISBN 3-540-63422-3

The volumes 3/I and 3/J present a modern account of polycyclic aromatic hydrocarbons (PAHs) and their heterocyclic analogs in the environment. The authors are internationally well-recognized scientists belonging to those working presently in the frontline of the different subfields of this interdisciplinary area of environmental science; they give an integrated thorough overview on this hot topic. Extensive cross-referencing between chapters provides the readers with an easy access to all major areas. Due to the huge amount of material the text is published in two volumes (3/I and 3/J). A major source of information and inspiration for all researchers actively working in PAH environmental chemistry or ecology.

Please order from
Springer-Verlag Berlin
Fax: + 49 / 30 / 8 27 87- 301
e-mail: orders@springer.de
or through your bookseller

Price subject to change without notice.
In EU countries the local VAT is effective.

Springer

Springer-Verlag, P. O. Box 31 13 40, D-10643 Berlin, Germany.

Gha.